Thirst

In Search of
Freshwater

Foreword by
Robert Macfarlane

First published in Great Britain in 2025 by Wellcome Collection

**wellcome
collection**

183 Euston Road
London NW1 2BE
www.wellcomecollection.org

Curator **Janice Li**
Editor **Ellen Johl**
Copy-editor **Linden Lawson**
Proofreaders **Sarah-Jane Forder** & **Clare Sayer**
Designer **Kay Ogundimu**
Producer **Petra Essing**
Typefaces **ABC Solar** & **Libre Baskervile**
Exhibition title design **Wolfe Hall**
Distributed by **Thames & Hudson**
Printed and bound in Great Britain by **Clays Ltd, Elcograf S.p.A.**

A catalogue record for this book is available
from the British Library.

ISBN 978-1-9998-0905-8

MP-7637.3/4200/04/2025/KO

Contents

Foreword

By Robert Macfarlane

A mile from my home a river is born. In a little wood between a railway line and a hospital, nine springs rise in a deep hollow of chalk. Their water pools, then flows as a slender stream into the city of Cambridge. There it joins the River Cam, which in turn joins the River Ouse, which in turn reaches the great estuary known as The Wash – where fresh and salt mingle, and river and ocean become united, not single.

The water of these springs is so clear that it sometimes seems not to exist at all. Like the 'well at the world's end' of which Neil Gunn wrote in his 1951 novel of the same name, it is so lucid that sometimes you must dip your hand into it to be sure it exists. Because it rises from deep underground, welling up from the aquifer which lies beneath the little wood, the springs' water is always a steady temperature of around 10°C. On bitterly cold winter days, the surface of the spring pool steams amid the frosted beeches and briars. On hot summer days, the pool is always cool – a place to drink from a cupped hand.

We know, in fact, from recent archaeological discoveries that people have come to these springs to slake their thirst since the Mesolithic period. Knapped-flint microliths and burnt hearth stones have been found in the field a few yards from the springs – and dated to between 6,000 and 9,000 years ago. Hunter-gatherers, Neolithic settlers, Iron Age farmers, Roman legionaries: they have all left their marks on the land around the springs, and all have been drawn here to drink – as have the aurochs, the deer, the wildcats, the hawks and the redwings. Thirst: an originary impulse and an ancient word – from the Proto-Indo-European *ters*, meaning 'dry'.

Tucked in the little wood, close to the springs, is a stone obelisk: 12 feet tall and made of polished grey-brown granite, with a sharpened tip. It is a surprising object to meet there. It has the air of an alien artefact, dropped from a spacecraft to embed itself, quivering, in the chalky earth. The obelisk commemorates the work of the city fathers who in the 1570s paid for the water of the springs to be conduited through the centre of Cambridge. This was a public-health initiative: the conduit was used to provide drinking water to the populace, to flush ordure and rubbish through the streets, and to reduce the incidence of disease.

The springs helped the city to live. Now the city threatens the life of the springs. Over-abstraction by water companies from the aquifer that feeds the springs, combined with climate-change-driven shifts to patterns of rainfall and drought, has left the flow at the springs perilously low. In 2019 an 'augmentation scheme' was installed at the little wood: a buried pump which, when groundwater levels drop, propels water from elsewhere in the catchment back into the springs to ensure they do not run dry. The springs' longevity is astonishing, but their fragility is extreme. In the little wood, a few hundred yards from one of the biggest hospitals in the country, the springs are on life support.

It has become customary in some circles to name the epoch through which we are currently living as the 'Anthropocene': the Age of Anthropos, in which human activity is the defining force, possessed of a terraforming power so great that it will leave a durable signature in the strata of the Earth – a signature legible to an imaginary geo-archaeologist who, millions of years in the future, will scrutinise the rock record and try to understand what happened long before.

Really, though, we are living through the Hydrocene – and always have been. When the first rains fell on this planet, more than 4 billion years ago, the land awakened. The beginning of the water cycle is presumed to have created the conditions within which life itself might flourish. Still that cycle turns. We all know the sequence, or think we do. Cloudbursts form streams, streams form rivers, rivers run seawards to become oceans, oceans evaporate into water vapour, and that vapour gathers in vast, invisible sky currents which flow upwards, counter-gravitationally, to condense and fall as rain and snow. *Da capo*.

The water cycle is now moral in dimension as well as geophysical, however. Today it is configured by power, violence and global capital, as well as by isobars, gravity and geology. It contains immiseration, injustice, hope, fear and possibility, as well as blizzards, glaciers and estuaries. Freshwater's uneven distribution is the driver of wars, migration patterns, global health trends – and even the planet's motion: the Three Gorges Dam in China has impounded so much water that it has measurably slowed the rotation of the Earth.

Astonishing new research has used satellite data and computer modelling to map the flows of every single river on the planet, for every single day over the past thirty-five years. This 'everything-everywhere-all-at-once' research frame reveals that human activity and the fossil-fuel-driven climate crisis, by accelerating glacial melt and shifting rainfall patterns, have caused almost half of the world's largest downstream rivers to lose both speed and volume of flow. Meanwhile, the smaller, steeper, upstream rivers are experiencing faster, more volatile flows and a dramatic increase in larger flood events.

Little of this is good. The slower-flowing bigger rivers mean there is less freshwater available in hugely populous areas, and less sediment is fortifying deltas against sea-level rise. Meanwhile, upstream catastrophic flooding is destroying homes, habitats and lives, human and more-than-human. The world's great glaciers and ice caps, which throughout the Holocene have acted as immense storage units for freshwater, are thinning towards depletion.

The crisis is one of imagination as well as of legislation. We have somehow forgotten that our fate flows with that of water – and always has. Our relationship with freshwater in particular has become increasingly instrumentalised and monetised: river understood as resource, not life force. The naiads and other water spirits – which in folklore figured the sacredness of individual springs, streams and rivers – have all been put to flight. Instead, water stocks are traded on futures markets, with disaster-capitalist investors gambling that crises to come will drive unit prices up. Water has become a liquid asset or, as Goldman Sachs hungrily described it, 'the new oil'.

How did it come to this? And what is to be done? In 1949, the American ecologist Aldo Leopold wrote an essay called 'Thinking Like a Mountain'. We might extend Leopold's thought experiment and ask what would it mean to 'think like a river'? To do so, it seems to me, would involve thinking both upstream and downstream in time, and recognising how every action is entangled in webs of inheritance and legacy. It would mean recognising the profoundly relational nature of being – that we are all water bodies, flowed through and flowing on. It would mean learning what the writer Barry Lopez called 'the syntax of the river'; that is, the infinitely ramifying connections between parts.

When we look into water, it lenses our vision. A refraction occurs: what appears to be the line of sight is re-angled slightly. This is referred to as 'parallax error' and we are told to correct for it. This is a misnomer, however. It would be better understood not as an error, but as a reminder that what we assume is often askew. Water, we might say, helps us see the world otherwise. It disrupts norms and presumptions.

Among those presumptions is that many of us have come, disastrously, to think of water as dead: as what Isaac Newton, in a letter of 1693, called 'inanimate brute matter'. This is a parallax error of immense consequence. We might reverse the terms of the proposition, and ask what it would mean to think of water as living, as alive? What would be the implications for our systems of governance, law and ethics, for our cities, communities and imaginations, if we were to recognise water bodies as beings with lives, deaths and even rights? These are radical questions, and very hard to answer. But even the asking of them is a start.

Rivers, streams, lakes and springs are easily wounded. A wounded river is one whose fish float belly-up, whose tributaries are rank with sewage, whose channel is choked with garbage. A dying stream is one who barely flows – who is close to 'deadpooling', to use the hydrologist's term of art.

Given a chance, though, water's life pours back. In the early autumn of 2024, the last of the great dams was removed from the Upper Klamath River, which flows out of Oregon and into California. Those dams had been built a century earlier and their construction had created impassable barriers to the migrating anadromous fish who for thousands of years had swum hundreds of miles up the Klamath, in order to spawn in its higher reaches. The removal of the dams – the largest de-damming project in US history – was the result of two decades of campaigning and activism, led by members of the Klamath Tribe.

On 15 October, within a fortnight of the removal of the final, immense 'Iron Gate Dam' just south of the California–Oregon border, something extraordinary happened. A sonar camera set up by scientists detected a single Chinook salmon migrating upstream past the pinch point where the Iron Gate Dam had stood. It was the first fish to make that journey in over a century, guided by an ancient navigational system and driven by an undeniable obligation.

I saw that sonar image the day it was released. It was grainy in the way sonar is – the fish just a blurry lozenge-line set against a blurry background. Definition didn't matter, though: the simple fact of it caused hope and joy to grow suddenly huge in my heart, and the heaviness of spirit from which I had been suffering for weeks that autumn began to lift. Strangely, I haven't been able to find the image again online, though I've searched several times for it. It has swum back out into the ocean of the internet. Perhaps I dreamed it. I don't think so. I know this for sure, though: water, once healed, heals us in return. ◆

Every

story

is   a   story

of

water

Natalie Diaz

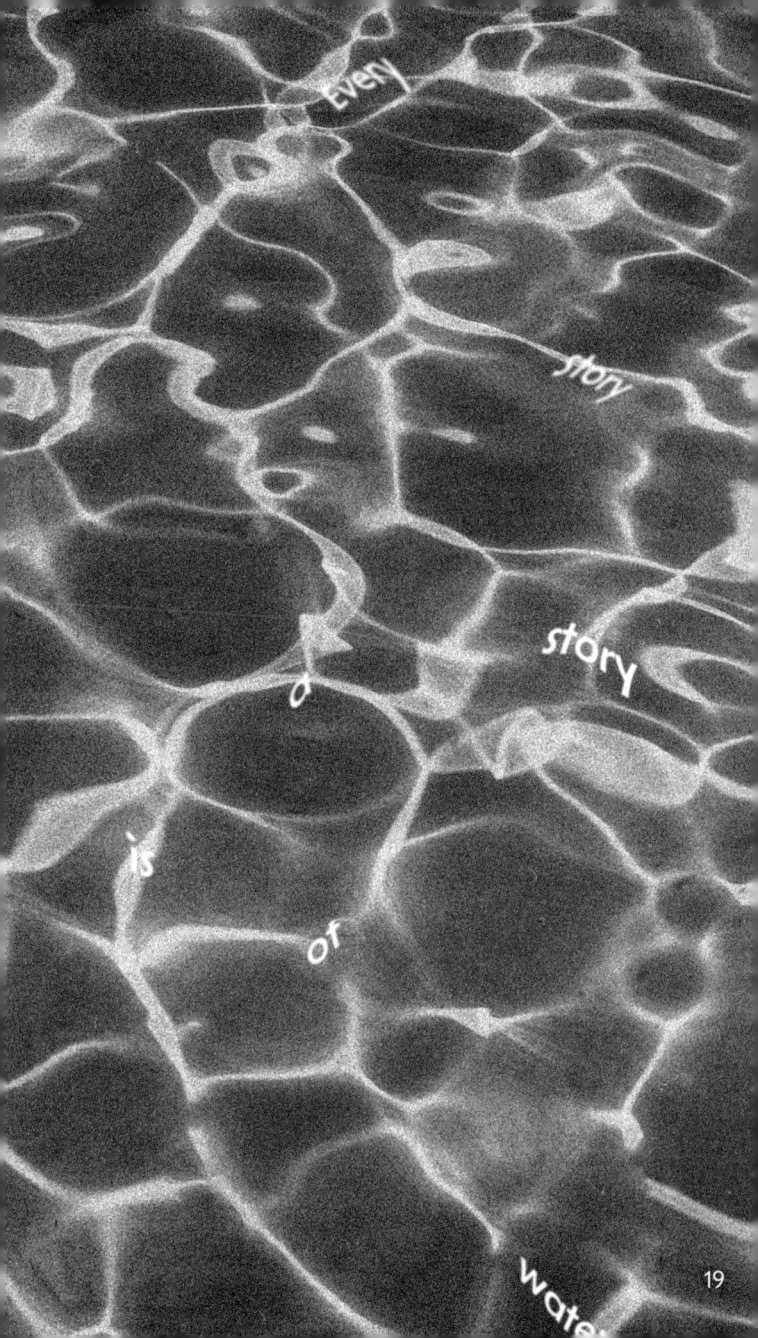

Every

story

story

a

is

or

wate

19

The world is blue at its edges and in its depths.
This blue is the light that got lost.
Light at the blue end of the spectrum
does not travel the whole distance
from the sun to us. It disperses among
the molecules of the air, it scatters in
water. Water is colorless, shallow water
appears to be the color of whatever lies
underneath it, but deep water is full of
this scattered light, the purer the water
the deeper the blue. The sky is blue for
the same reason, but the blue at the
horizon, the blue of land that seems to
be dissolving into the sky, is a deeper,
dreamier, melancholy blue, the blue
at the farthest reaches of the places
where you see for miles, the blue of
distance. This light that does not touch
us, does not travel the whole distance,
the light that gets lost, gives us the
beauty of the world, so much of which
is in the color blue.

For many years, I have been moved by the blue at the far edge of what can be seen, that color of horizons, of remote mountain ranges, of anything far away. The color of that distance is the color of an emotion, the color of solitude and of desire, the color of there seen from here, the color of where you are not. And the color of where you can never go. For the blue is not in the place those miles away at the horizon, but in the atmospheric distance between you and the mountains. ◆

Rebecca Solnit

Remem

bering

By Elif Shafak

Water remembers.

It is humans who forget.

For so many years now I have been intrigued by and obsessed with water. We tend to take it for granted, but water is perhaps the greatest mystery in our lives. There is still so much we do not understand or know about it. The climate crisis is essentially a freshwater crisis. That feeling of impending catastrophe is especially acute in the part of the world where I come from. Out of the ten most water-stressed nations, seven are in the Middle East and Africa. The legends, stories and silences of Mesopotamia are shaped by rivers.

Like buried rivers, inherited traumas flow unseen beneath the structures we have carefully built, always threatening to overflow their banks. Unless we remember, we cannot repair, and what we cannot repair will keep resurfacing and coming back, again and again.

As human beings, our relationship with nature today is mostly based on consumption. We have created a system in which we have turned ourselves into consumers of water, consumers of air, consumers of earth without realising that we are neither above nor outside nature. In truth, we are only a small part of a delicate and complex ecosystem. I sincerely believe we need to rethink our entire connection – or lack thereof – with the environment.

Our rivers are dying. This has massive consequences – political, social and economic. It also affects women more deeply. Women are water-carriers. When there is no water, women must walk longer distances and, often along the way, experience more gender violence. We tend to put these issues in separate boxes: 'water scarcity', 'gender inequality' or 'the loss of culture', but in reality, everything is connected.

◆

Water

The struggles of minorities and Indigenous peoples for water rights

By Vandana Shiva

Justice

In Urdu and Hindi, water is known as *ab*. The word *abadi*, derived from *ab*, is also the word for community. People come together to form communities wherever there is water. *Abad raho* is used as a greeting to encourage prosperity and abundance.

Water is a living cycle that connects biosphere and atmosphere. Forests, rivers and oceans are interwoven by the water cycle, keeping life on this planet in a state of constant regeneration, renewal and recycling. Water is also the flow connecting life and people. Simply put, water is the basis of life-affirming cultures.

Sustainability and justice are part of the same ecological process. We are all Earth citizens. We all participate in the water cycle. Our bodies carry water and make us all participants in the endless movement, flow and distribution of life across the planet.

Whether a community faces scarcity or abundance depends on how we relate to water and how water is distributed. Economies and cultures that disregard the water cycle and waste, pollute or destroy water are generating scarcity and injustice even in places where there is abundant ground and surface water. Those that save every drop can create abundance out of scarcity.

When the rich, powerful and dominant economic forces of society take more than their Earth share, ecosystems, plants, animals, women, Indigenous communities and minority groups are deprived of their share of water for life and livelihoods, leaving entire communities to carry the heavy burden of water poverty. Women in communities all around the planet fetch and carry water on a daily basis. Extraction and pollution of local water systems increases their burden.

In 2005, India's National Commission for Women asked the Research Foundation for Science, Technology and Ecology to conduct a report on women and water. Our participatory research, conducted with rural women across India, showed that if opportunity costs were considered, the cost of fetching water is almost equivalent to 160 million working days each year, which translates into 10 billion rupees.

Women are the first to know when water systems are being destroyed by greed.

In the 1970s, women of the mountain regions of the Himalayas rose to form the

Chipko movement. Wide-scale deforestation had led to the destruction and disappearance of water springs, destabilisation of the mountain slopes and major landslides. *Chipko* means to hug. Women decided they would hug the trees to stop destruction. Chipko women reminded the world that forests are not timber mines for the extraction of profits. Forests are the sources of water, soil and oxygen.

In 2002 women from a small village in Plachimada, Kerala, started a *satyagraha* – an act of non-violent civil disobedience aimed at shutting down a Coca-Cola plant that was extracting 1.5 million litres of groundwater per day, causing water famine in a water-rich region. By 2004 the women had shut down the Coca-Cola plant. As a result of their efforts, groundwater was recognised as a commons for the local community to enjoy.

In Bosnia and Herzegovina, a group known as the Brave Women of Kruščica led a 503-day blockade of heavy equipment that resulted in the cancellation of permits for two proposed dams on the River Kruščica in December 2018. The Balkans are home to the last free-flowing rivers in Europe.

These and many other examples around the planet show that women not only carry water to supply households, but also tend to lead the struggle to defend, protect and fight for the life of waters in the face of rampant private ownership and extraction.

The privatisation of water, imposed on the Global South in recent years, is a litany of tales of corporate greed.

One of the most infamous of these occurred in Cochabamba, Bolivia, in what is now known as the Cochabamba Water War. In this semi-desert region of the High Andes, water sources are vulnerable and precious. In 1999 the World Bank recommended privatisation and a monopoly over Cochabamba's municipal supply through a concession to International Water, a subsidiary of the US-owned Bechtel Corporation. In October 1999 the first water bill was passed, granting the privatisation of Cochabamba's water. Privatisation led to such high prices that people in this deprived region found themselves spending up to 20 per cent of their monthly wages on water bills.

In January 2000 a citizens' alliance called Coodinadora de Defensa del Agua y de la Vida (Coalition in Defence of Water and Life) was formed. The alliance shut down the city for four days through mass mobilisation. A Cochabamba Declaration was subsequently issued.

Protests continued until the privatisation law was annulled in April 2000. Contracts were terminated, and new water-resource laws were drafted following citizens' participation.

In India, the World Bank imposed a water privatisation project in Delhi in 2001, only months after the Cochabamba Declaration. The consultants appointed for this project were Price Waterhouse Coopers. Waters from the sacred River Ganga were to be privatised and sent to the Suez-Degrémont water plant at Sonia Vihar. A water democracy movement soon came together to connect the people displaced by the Tehri Dam and farmers whose fields are irrigated by the Ganga Canal in the slums of Delhi.

On 8 August 2004, on the eve of Quit India Day, which marks when British colonisers were told to leave India, we gathered on the banks of the Ganga at Haridwar. We were joined by Sunderlal Bahuguna, the leader of the Chipko movement, Rajendra Singh from Rajasthan and Oscar Olivera from Bolivia, one of the leaders of the Cochabamba movement during the Cochabamba Water War.

We took a pledge that we will never let the River Ganga be sold to any multinational company. Ganga is revered as a mother – Ganga Maa. We will never allow our mother or her water to be sold to Suez-Degrémont or any other corporation.

We carried out water pilgrimages, staged water protests and held public hearings. On 9 August 2002, 5,000 farmers gathered in a rally. In 2004, 150,000 people joined a march in the heart of Delhi, signalling a commitment to keep the waters public. The movement forced the cancellation of the World Bank project.

Water is life, not merchandise.

By reclaiming water from corporations and the market, the citizens of Bolivia, India and other countries have illustrated that privatisation is not inevitable and that the corporate takeover of vital resources can be prevented by people fighting for water democracy at a planetary scale. The role of minorities and Indigenous peoples in this fight for water justice cannot be praised highly enough.

The examples from Bolivia and India are not isolated ones. The unjust privatisation in water-stressed countries is a global pattern of grotesque proportions. This is far from over. In fact, the struggle to secure water justice for communities, especially those belonging to minorities and Indigenous peoples, has intensified all around the world.

According to a recent UN report on the state of the world's water, 'over the last few decades, the water crisis has deepened on a planetary scale. 75 per cent of available freshwater is now used for chemical- and water-intensive agriculture, which also leaves rivers and lakes polluted with nitrates and pesticides.' According to the same UN report, more than 5 billion people could suffer water shortages by 2050 due to climate change, increased demand and polluted supplies.

If water is put on the open market like oil and gas, it will inevitably lead to rising prices in a world desperately in need of water for life.

Ten years ago, the UN General Assembly adopted a resolution recognising that clean water and sanitation are basic human rights. The move to commodify water on Wall Street directly threatens these human rights and puts billions of people in jeopardy.

We demand that people everywhere – and all governments – reject the commodification and financialisation of water, and that water is finally recognised as a public trust as well as a human right in law and practice for all time. ◆

6

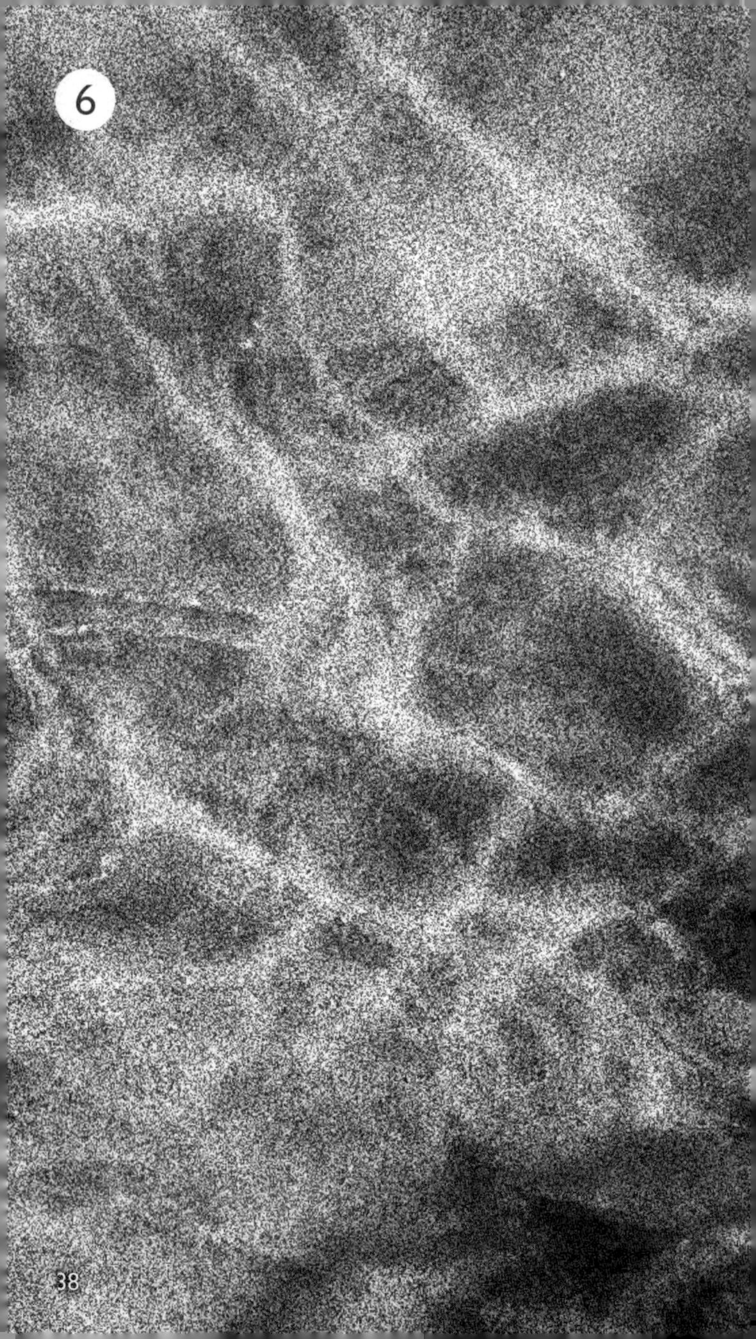

The

Secret

By Emmanuel Vaughan-Lee

We

Thirst For

Growing up, I was taught a Sufi story about a school of fish who went in search of Water. They had heard whispers of this sacred substance and it intrigued them; but none of them had any idea what it was, so they started asking other fish as they swam by. No one could tell them. Then one day they came across a fish who said he had heard of a great, wise old fish who lived far away at the end of a lake. So they all set out in search of this wise one who could tell them about Water.

I often think that we are like these fish, seeking out something that is all around us – that is most essential in life – but that we have become blind to. And what is more essential than Water? But rather than whispers of a sacred substance drawing us on a quest to find a wise old fish who will enlighten us, it is the cry of the Earth that is trying to get our attention. It wants us to remember the true nature of Water.

And what is the true nature of Water?

It is the most foundational element that flows at the border of spirit and matter. It is a substance that purifies, sanctifies and underpins all life; that was once revered as river goddesses and divine oceans. Is this what we thirst for, what the cry of the Earth implores us to remember, like the whispers heard by the fish?

Sufis say that the vessel may change, but the Water stays the same. Symbolically, I believe this to be true, but we have changed Water so fundamentally, polluted and abused it to such a degree, that it has almost become something else. It is now a reflection of our human-centric way of life, which has forgotten the secret of this substance; forgotten that it flows at the border of spirit and matter – for it is only there that the true nature of Water will reveal itself once again.

That is now the task: again to remember this secret, to remember that the worlds of spirit and matter must flow together, to remember that this is what we truly thirst for. ◆

Hawthorn

Healer of Hearts

By Lora Aziz

يا زعرور يا امور
اللى في سينا مشهور
من الجبال والبحور
يا زعرور يا امور

Hawthorn, you beauty,
Famed across Sinai's land,
From the mountains to the sea,
Oh Hawthorn, you beauty.

October 2021. While resting at Al'Karm – an ecolodge cradled at an entrance to the Wadi Gharba Valley, the historic pilgrims' route to St Catherine's Monastery in South Sinai – I dream of hawthorn, زعرور. Its ruby fruits, swollen like cherries, burst on my tongue, their juice pooling and running down my chin. It is as if the tree itself is reaching out to offer its essence, inviting me into its ancient, tangled embrace

The hawthorn calls to me, its presence quietly unfurling in my dreams. I feel its pull, subtle but insistent. Determined to find it, I ask around. A friend in Wadi Shreij, not far from St Catherine's Monastery, welcomes me with coffee brewed with *mistika*. From her veranda, pistachio, almond and lote trees frame our view, their branches stretching towards Gabal Safsaf, Willow Mountain. Bathed in a golden, sandy light, the mountain seems to look back at us knowingly. She points towards it: 'It's up there ...'

The mountain is alive with whispers – ancient gardens with crumbling borders, dusty churches standing as sentinels of something timeless. The air is filled with the mingled scents of wild thyme and wormwood. We walk slowly, our steps deliberate, the mountain pulling us into its rhythm. And then, the hawthorn.

A few days later, the hawthorn calls to us again, on that mountain – its thorns fierce and strong, like guardians of its secrets. Its red, brittle bark flakes away, as though bearing too much of the mountain's memory. Late-autumn sunlight catches the berries, glinting like desert jewels, ready to be shaken loose. Beside the tree stands an old well, its cool, mineral-rich water a lifeline to the arid land. Wells are sacred in the desert, holding more than water; they hold memory, sustenance, connection.

But the long-term decline in rainfall has caused the water table to drop, leaving many wells dry. In this fragile landscape, knowing your water source is essential. To understand our water source is to understand life itself, for water binds us to the Earth and our bodies, sustaining everything we are.

After several hours of walking we sit under the hawthorn's shade, brewing tea with water from the well. As we wait for it to boil, I feel my breath returning to its natural rhythm. In the stillness I run my hand over the bark, feeling its scratchy surface on my skin – a direct connection, tangible and grounding. I have arrived.

I shake the tree and its red berries rain down, scattering across the rocky ground. I fill my bag with some small fruits of medicine – hawthorn, the healer of hearts, representing blood and circulation. The water within us that moves like rivers beneath the skin.

Hawthorn Water Essence Recipe

A simple recipe that captures the essence of hawthorn, offering a moment of embodied connection. This humble tree, thriving in St Catherine's arid embrace, is also familiar to the UK, where it lines hedgerows and marks the boundaries of the commons. In folk medicine in both lands, hawthorn is regarded as the ultimate heart tonic, known to enhance circulation.

Ingredients

- A handful of fresh hawthorn berries (gathered with gratitude, ensuring the tree is left abundant)
- Fresh, clean water (find your source)
- A small glass jar or bowl
- Sunlight or moonlight

Method

1 Place the hawthorn berries in the jar or bowl, covering them with water. As you do so, offer a moment of gratitude to the tree and the land.

2 Leave the jar under the tree you gathered from, in sunlight for a day or moonlight overnight. Let the light infuse the water with the vitality and memory of the hawthorn.

3 Strain the berries and store the water in a clean jar. Keep it in the fridge and use within a few days.

This essence can be sipped in small amounts, added to tea, or used to honour the heart and the water that flows within and around us. It is a gentle reminder that the gifts of the Earth, like the hawthorn, are always with us, inviting us to listen, to feel and to heal.

How sweet. That rain. How something that
lives only to fall can be nothing
but sweet. Water whittled down to
intention. Intention into nourishment.
Everyone can forget us – as long as
you remember.

Ocean Vuong ♦

Ancient

Green

By Robin Wall Kimmerer

One summer day in Alaska, I stood within a glacial cave, blue and strange beneath the ice. I heard the plunk of drips falling into the meltwater pool and shivered in the cold blue light. I listened to the calls of ice becoming water. There's a story that begins here, or maybe it ends. It depends on us.

The last time the glaciers melted here in the Adirondacks, they left this boulder field behind. Hundreds of glacial erratics – scraped from the ancient Laurentian Shield, rolled here beneath the ice sheet – dot the landscape. Today their scarred granite surfaces are robed in mosses. The air itself is charged with their radiant green. The boulders look like a herd of ice-age musk oxen, frozen in place with a thick coat of green fur, grazing beneath a post-Pleistocene canopy of birches, maples and hemlocks. As a bryoecologist, I've spent decades observing these islands of mossy rock. What seems like a lifetime to me is barely an eyeblink in the 10,000-year rest that allowed these rolling stones to gather some moss. Mosses and rocks take the long view.

Mosses, I think, are like time made visible. They create a kind of botanical forgetting. Shoot by tiny shoot, the past is obscured in green. That's why we have stories, so we can remember.

The mosses remember that this is not the first time the glaciers have melted. If time is a line, as Western thinking presumes, we might think this is a unique moment for which we have to devise a solution that enables that line to continue. If time is a circle, as the Indigenous world view presumes, the knowledge we need is already within the circle; we just have to remember it to find it again and let it teach us.

Since the momentous colonisation of land 450 million years ago, when the first moss set leaf on rock, everything on Earth has changed. All those species, entire phyla – gone. And yet the mosses are still here, their contemporary form indistinguishable from their fossil ancestors. They have drunk from the fountain of youth, or maybe the fountain of longevity, flourished beneath a sky of pterodactyls, and flourish today under a sky of weather satellites that tell us the oceans are rising and the ice caps are melting.

All things pass away. Oh, lovely, cool shaded maples, towering pines, waving grass and extravagant lilies, will you too pass away in this overheated greenhouse, yielding to the ones who are yet to come?

They do not discriminate in their coverage, be it a post-glacial boulder or a car long abandoned in the woods – all are blanketed. I once found a pair of logger's boots on a cut stump, robed in moss, with sporophytes rising through the eyelets. In their vibrant verdancy they seem to say, where there is light and water, life will win.

They cover the inanimate with the animate. Without judgement, they cover our mistakes, with an unconditional acceptance of their responsibility for healing.

They will cover the abandoned frack pads with the same tenderness as the bare rubble of a melted glacier. Mosses were the first plants to blanket the Earth. I wouldn't be surprised if they are also the last.

It doesn't have to be that way. What if we look at the mosses not only as healers of land, but as teachers of how we might live?

I don't know about you, but in this moment on the cusp of climate catastrophe I long for a wise elder, a teacher to guide us. At the time of the sixth extinction, might we stop wringing our hands long enough to sit quietly at the feet of the ones who have avoided every era of extinction since the dawn of life on land?

I've had the privilege of being a student of mosses for most of my life, kneeling before them, writing the stories they have shared with me. It never gets old, peering into moss rainforests where the trees are just an inch tall and brightly coloured mites perch like parrots on their lustrous leaves. Each one as different

Ancient Green

from the next as a palm tree is to a magnolia, their beauty draws me back again and again. There is no light like moss light after a rain shower, when they glow and glisten, water beading up on intricate leaves smaller than a raindrop. And the smell ... the woodsy, humic richness that reminds us where we came from.

Defying evolutionary expectation of extinction, they have come through ice ages, eons of warmings, dryings, shifting of continents, uplift of mountain ranges, the rise and fall of countless other beings, from *Tyrannosaurus rex* to *Homo sapiens*. They have lasted.

The needs of a moss are simple and not unlike our own: food energy, water, warmth, a place to raise their young – and beauty. But their means of meeting those needs are very different.

Mosses make minimal demands on their surroundings. All they need is a little light, a sheer film of water and a thin decoction of minerals, delivered by rainwater or dissolution of rock. If they are hydrated and illuminated, they will exuberantly photosynthesise and expand the green carpet. But when times are tough, most simply stop growing and wait until water returns. They don't die, they just crinkle up and pause, following the rhythms of the natural world, growing in periods of abundance and waiting through periods of scarcity: a wise strategy for life that is in tune with uncertainty.

Moss lifeways offer a strong contrast to the ways we've organised our society, which prioritises relentless growth as the metric of well-being: always getting bigger, producing more, having more. Infinite growth is ecologically impossible and exceedingly destructive, as it demands the transformation of the lives of other beings into raw materials to feed the fiction. Mosses show us another way – the abundance that emanates from self-restraint, from enoughness. Mosses have lived too long on this planet to be seduced by the nonsense of accumulation, the delusion of permanence, the endless striving for productivity. Maybe our heartbeats slow when we sit with mosses because they remind us that contentment could be ours.

The biggest factor that limits their growth is water, since they can only photosynthesise when wet, which is why moss is so lush in waterfall splash zones and dripping temperate rainforests. But even in the driest desert there are mosses, living on dew. Without the fancy water-conservation mechanisms of more advanced plants, mosses rely on an intimate relationship with water droplets to thrive. Their thin leaves, just one cell thick, overlap one another like shingles, and each minuscule leaf is sculpted with nubs and grooves and frilly hair-like extensions to hold a film of water by capillary action. The whole architecture of a moss fosters the love affair between leaf and water, the physicochemical attraction of water for cellulose, in order to hold water close.

Watch a raindrop land on a dry moss and you might learn something more about living well. The water seems to move of its own accord, running along the leaf surface and climbing up to a branch tip, defying gravity through the affinity between moss and water. Water is moved not by clanking pumps and pipes, but by the sculpted shape of the plant. The architecture of a moss is designed to move water without expending any additional energy at all, rather by simply harnessing the forces of attraction between water and cellulose. Such economic elegance requires accepting natural forces and letting them shape your way of life. I like to imagine a human community designed the same way, embracing natural forces rather than obstructing them.

Water is held best not by an individual shoot, but by the collective sponge of an entire colony. Competition as an organising economic principle has fairly predictable results: the rich get richer and the poor get poorer. But mosses organise themselves for a different economic outcome: shared wealth. Rather than competing for scarce water, a moss is designed for equitable sharing. Water is passed from shoot to shoot across leafy bridges and down canals of capillary space to moisten the entire colony, not just an individual. Ecological rules usually dictate that crowding is deleterious, but mosses break those rules. A community of mosses can gather and retain precious moisture much more effectively than a lone individual. We know this kind of mutual support instinctively – in times of trouble people leave their isolated lives and band together. But we forget.

This is the environmental philosophy of mosses, that small is beautiful. They remind us of the virtue of humility, a value in short supply among the people of the Anthropocene. This view is hard for humans to accept, with our love of power and stature.

We humans pride ourselves on living by the rule of law, but the laws we choose to obey are only those of our own making. We ignore ecological laws as if the fiction of human exceptionalism meant that thermodynamics did not apply to us. Whether we choose to heed them or not, natural laws will prevail. Arrogance has brought us to the brink. The laws of nature will bring us to our knees. And then perhaps we will see the mosses. ◆

Lake

By Jessica J. Lee

Lights

1. Fish-scale silver
Lake Stechlin

The shift begins the moment the forest clears and the horizon stretches ahead of me. The sunlight is intensified by cloud, that curious kind of winter sky, gleaming bright white. The lake beneath it is a mirror, glass-ice doubling the light. I reach a gloved hand to my brow to create shadow; it is winter and I've grown unused to anything but darkness.

My eyes adjust, and as I trace the line where the lake meets the forest on the opposite peninsula, I see the illusion break down: there is a dark line on the distant shore where water glows black. A crack in the reflection. At my feet, on this bank, the colour is different: pale-brown sand and chipped white ice, a yellow froth where the water moves unfrozen.

I often think you can tell the temperature of a lake by looking at it. Greys and whites are almost always frigid; the clearer, the colder. Swim enough and you start to think you know the water, that it isn't indifferent to you.

But this morning, as I slip off my winter coat and woollen socks and step barefoot into the water, I sense the vastness of Stechlin for once. The way it resists my understanding. The lake dips round a bend on the horizon, half its body swathed in mist. This lake was once said to be among the clearest in Germany. Today the lake is almost entirely frozen, save for this small pool – hardly an invitation. Silver swells around my waist and I watch small ripples rock the ice ahead. Pain and joy flood me in the cold. And this sensation is why I came. Afloat in the body of the lake, I watch my limbs distort, the water gleaming, sharp as stone.

2. Petrol-green
Schmaler Luzin

I am always wanting more from water: more sensation, colour, light. I've been swimming the lakes north of Berlin for the better part of a decade, and in that time have come to know each water's peculiarities: the ferrous glow at Mechesee, the emerald gleam at Liepnitzsee. In Feldberg I could distinguish the lakes of a glacial chain by texture and by hue, from weekends paddling the length of them, evenings wading on their shores. But my swims have often been short: more to feel the water than to get anywhere in particular. It was about my body, what the water helped me feel. In ten years of swimming, it is as if I've merely sketched each lake, like there is much I haven't known.

This weekend I am here to swim: not

simply to languish in the shallows, but to really swim. My friend Becky has invited me to join her trekking group, distance-swimming the lake chain over the course of many days. This morning I was given a flame-orange tow-float and a glance at a paper map: two routes I can take, depending on my energy. Both trace the northern length of Schmaler Luzin, a skinny finger-lake surrounded by thick forest.

The other swimmers here are stronger than me, more acclimatised to distance. Two of them, someone whispers, are retired Olympic athletes. My usual breaststroke won't serve me well, so I've promised myself I can simply try the shorter route – 3 kilometres – with peppermint tea and chocolate at the end.

The group splashes out ahead of me with swim caps and goggles pulled tight. In the moments it takes me to ease myself into deep water and begin to find my rhythm, they've all crawled ahead, save a single swimmer to my left. Amanda. Her elbow slowly rises and falls with each stroke, and every so often she slows, turns over and looks around. This morning during breakfast she told me that she is here to notice things: the birds, the light, the sky. To get a feel for the water. I nodded in agreement. I am glad, at least, to have this water to ourselves.

The middle of Luzin is deep, and with all the other swimmers far ahead the surrounding lake is wide open. It is a cool day, summer light shaded by grey clouds, and every so often the sky above me shifts. I am swimming more steadily than I ever have before, my face sub-merging with every stroke, my eyes tracing the forest with every breath. I feel my legs warming, my arms moving without thinking about it.

All the years I've swum these lakes, I've never really seen it like this. The way colour changes in the deepest middle, the way the eye focuses on colour when there is nothing else to do but swim. I've swum here for sensation. But swimming distance, with duration, my body has all but disappeared: I am only eyes and breath. My skin exists only when channels of cold water rush over me.

The swim takes eighty minutes at a slow pace. To pass the time I name everything I see: petrol-green pool flecked with gold spots. Whaleback-grey waves when clouds huddle over. Obsidian and white on a swallow swooping low. Green-black blur of summer trees on the shore. Every few minutes, I turn left to look: Amanda still swimming, pale arms pulling towards me. Orange floats on the surface, trailing us both.

3. Periphyton-gold
Lake Stechlin

I remind myself that there are other ways to know a lake. Ways to move beyond swimming out and back at the surface, seeking pleasure in my body, seeking cold and pain and elation. I begin to tell myself swims are not enough. That my relationship to a body of water shouldn't be so transactional.

I read about Stechlin's history, make visits to the scientists working on its shore. They plunge Secchi discs from the sides of boats and record the depths at which they disappear: where clarity recedes. They teach me how this lake, once famous for transparency, is growing dim and clouded with each season. The water is warming in all the lakes. There is too much nitrogen here, seeping from the ground, and what was once nutrient-poor water is filling with clouds of green. The photic zone – where visible light reaches – is changing in this place. I decide then that I need to see it: to see beneath the surface to the lake light that is waning.

I spend my weekends training to free-dive with a teacher who knows this place. I learn to equalise the pressure in my ears, to soften muscles, slow my heart. One Saturday I spend dead-man floating in the shallows of the lake: breath held, focus narrowed on gold-red grains of sand. I feel the light closing after two minutes, tunnel-vision setting in. I breathe.

When I am ready, my teacher marks a point far distant from the shore. We snorkel there and drop a dive line, 10 metres down. 'Bereit?' he asks. I raise my thumb and roll over to my front. Ready.

The water around me is black at the surface, but the second my face goes under, the light changes. I'd imagined shafts of gold reaching clear into the deep, but what I see instead is a diffuse glow, toffee-softness over blue. I duck-dive down, tracing the course of our orange rope. It looks white underwater, a thin band of crisp light where everything else seems clouded. My arms push outward, my legs submerge, and my right hand drifts to my nose. I follow the line: past the tape marking 4 metres – everything still light – and the tennis ball marking 6.

And it is then that I can see it: a lake I once believed silver-clear now filled with dull colour at the bottom. Each branch of milfoil growing here is coated, as if in furry, unpolished gold. Periphyton – a mixture of algae, cyanobacteria and other detritus – grows where the lake's balance has been disturbed. I see it cover every submerged surface – like snowfall or fallen ash.

I do not know how many seconds have passed. The dive line drops beneath the gold cloud, light receding around me. There is pressure in my ears and throat, my lungs demanding air. I cannot go further today – my body is exhausted, and I still need to swim back to shore. I know only that I must return. But I had wanted to witness this lake changing, and for just a moment, I am here. ◆

A river passing through a landscape catches the world and gives it back redoubled: a shifting, glinting world more mysterious than the one we customarily inhabit. Rivers run through our civilisations like strings through beads, and there's hardly an age I can think of that's not associated with its own great waterway. The lands of the Middle East have dried to tinder now, but once they were fertile, fed by the fruitful Euphrates and the Tigris, from which rose flowering Sumer and Babylonia. The riches of Ancient Egypt stemmed from the Nile, which was believed to mark the causeway between life and death, and which was twinned in the heavens by the spill of stars we now call the Milky Way. The Indus Valley, the Yellow River: these are the places where civilisations began, fed by sweet waters that in their flooding enriched the land. The art of writing was independently born in these four regions and I do not think it a coincidence that the advent of the written word was nourished by river water.

Olivia Laing ◆

12

Rivers

of

Change

By Joycelyn Longdon

Systems of oppression are holding our individual and collective imaginations hostage. They have always led to the disconnection of myriad communities – but especially Indigenous and marginalised ones – from their Earth-honouring cosmologies. We need beliefs and practices that are rooted in the knowledge that all beings on the planet – human and non-human – are relations, not resources. The claws of colonialism and capitalism that scar the land, our lives and our souls, work to seize all traces of animistic acknowledgement, leaving us untethered, unable to ask whether the ocean cries for its corals or a glacier mourns its death.

Does a glacier mourn its death? When we ask these questions, we acknowledge that, in order to live better with the planet and each other, we must come to see ourselves in every river, rock and rainforest. These questions help us see the living world not as a passive backdrop but as a complex, dynamic and vastly intelligent being that has so much to tell us about where we have been and where we are going.

What thoughts at the source will trickle and drive rivers of change, transforming rivers and landscapes and minds and hearts when we ask whether glaciers mourn their death, whether trees miss their fallen kin, or what the time-travelling rivers whisper to the salmon and the sedges and the willow and the otter?

As a researcher working at the intersection of technology and environmental justice for tropical forest conservation, I harbour a bias in thought and advocacy towards the terrestrial, with less engagement with freshwater than I would like to admit. Yet, on deeper reflection, I find that – whether in the forest in Ghana, hopping over and stopping for a refreshing drink of crystal-clear water from the many rivers and rivulets, or in London, spending dazzling evenings strolling along the Thames, or in Lyme Regis, braving the October sea to splash and play with my best friend, or in Cambridge, inflating blow-up kayaks with my husband for an evening paddle, or in Cornwall, shaking off the nerves before jumping from the cliffs into Boscastle Harbour – I am constantly held, served and witnessed by water. These fluid, reflective, metallic, soothing, energetic bodies bear witness to our lives. The question becomes, how earnestly, how deeply, do we bear witness to theirs?

In March 2023 for World Water Day, Naveeda Khan, Associate Professor of Anthropology at

Johns Hopkins University, wrote an essay reflecting on how efforts to grant legal personhood to rivers 'provides a conceptual bridge for the imagination to take flight to explore other possible relations to rivers', beyond dumping sites on the one hand or 'ecological services infrastructure' on the other.[1]

The movement to fight for and achieve legal personhood for rivers across the planet is awe-inspiring, to say the least. Through various legal claims, rivers such as Whanganui in Aotearoa (New Zealand) or the Amazon in Colombia have gained personhood, acknowledging and formally honouring our interconnection with and dependence on non-human life within the law. It is a Western formalisation of a truth that is indisputable to many Indigenous communities: that the river is our ancestor, it is our kin. While not a perfect solution (does one exist?), it is but one example of how we can, and already do, collectively reimagine systems away from domination and extraction and towards kinship.

However, the work of reimagining must not be limited to global systems of politics and policy alone. It must, importantly, become a collective practice, bringing hearts, minds and hands together locally, as well as globally. We must not neglect the reimagination required to restore our local ecosystems and our individual and shared relationships with them.

In the autumn of 2024, I was incredibly proud to have been appointed as trustee for Action for Conservation (AFC), an environmental charity that works with diverse groups of young people aged twelve to twenty-four across the UK. Its mission is to bring the magic of nature to young people's lives in order to inspire and empower them to become the next generation of environmental leaders. One of the many reasons why I was so excited to join the team was because of the organisation's reimagination of the traditional approach to nature restoration projects: shifting focus instead to intergenerational ways of working, where young people are at the heart of designing and delivering large-scale, long-term restoration alongside landowners, farmers, ecologists and local communities. An incredible example of this practice can be seen through the Penpont Project, the UK's largest intergenerational nature restoration initiative.

Penpont is a 2,000-acre upland estate located in the Upper Usk River Catchment in the Bannau Brycheiniog (Brecon Beacons), Wales. The River Usk, which weaves through the gardens of Penpont, providing a beautiful picnic spot for visitors, is an essential wildlife corridor and key breeding area for nationally and internationally important species.[2] Yet those who live with and love the Usk worry that it will become known as 'the river that was lost'. Along with the River Wye, the Usk has the highest levels of phosphate in Wales,[3] and as Phil

Waggott, a local who has been fishing on the banks of the Usk since the 1980s, stated clearly, 'is full of sh**'.[4]

The Penpont Project is just one of many campaigns working to restore the Usk and its surrounding ecosystem, but it embodies what is, unfortunately, a unique perspective and ethos in British conservation. Theirs is an approach that not only honours the aliveness of the living world but is rooted in the knowledge that we, too, are a part of, not above or outside of, the ecosystems that sustain us. As written in the project documentation, the objective is to:

> achieve not just ecological restoration, but bio-cultural restoration – a process by which people and land recover diversity and resilience together within a place. This means seeing ourselves as part of nature acting with other species, not as external agents acting upon the living world.

Aiming to build an alternative model, away from more common attempts to protect and restore nature in the UK that overlook, ignore or marginalise ordinary people – especially those with the greatest stake – AFC introduced and facilitated a process of eco-cultural mapping. Sometimes referred to as 'social cartography', eco-cultural mapping combines flows of information from science with the traditional knowledge of local people and 'outsider' perspectives. By mapping, through three participatory phases, the past, present and future of the landscape, a collective action plan was devised in the form of a 'Rivers of Change' visualisation.

The mapping process was intergenerational and interdisciplinary, bringing together young people, landowners, tenant farmers, local historians and ecological experts, and scientists. Taking rivers themselves as a visual reference, the visualisation laid out the many streams of work and action that would take them from the present day to their collectively imagined future of the landscape.

What if we expanded this practice? Mapping our collective human and non-human histories, presents and futures? Drawing ourselves as the river and all beings within its catchment? Committing not just to rewilding but to biocultural restoration across all our precious waterways and water bodies?

A daunting task. A task that requires the patience, fluidity and transformative nature of water itself. But it does not have to be complicated. Maybe all we need to begin is to carve out time to sit riverside, and ask the water where it has been, how it lives now and where it wants to go. In its true nature, the river will reflect our questions back to us. Asking each of us:

Where have you been?

How do you live now?

Where are you going? ◆

Notes

1 'Legal Personhood of Rivers and the Failure of Imagination', https://dukeupress.wordpress.com/2023/03/22/legal-personhood-of-rivers-and-the-failure-of-imagination-a-world-water-day-guest-post-by-naveeda-khan/.

2 https://naturalresources.wales/media/662000/SSSI_1232_Citation_EN001606d.pdf.

3 https://beacons-npa.gov.uk/environment/love-our-rivers/#:~:text=The%20Rivers%20Wye%20and%20Usk,but%20we%20need%20your%20help.

4 https://www.walesonline.co.uk/news/wales-news/wales-most-protected-river-actually-29879139.

13

The

Parable

of

By
Raqs Media Collective

the

Step Well

A thousand poems will tell you that a woman goes to the water's edge, to the threshold of a well, to the riverbank. She walks, reckless, under cover of night, in the stillness of the afternoon, in storm, in rain, in scorching heat, to wait for her lover. They even have a name for her – Abhisarika – the one who wanders.

Waiting is the first act of desire.

Love is a kind of thirst. But is thirst a kind of unrequited love?

The hunter and the hunted, the lion and the lamb, go down to the waterhole. Thirst levels the playing field.

What comes first, the thirst for water, the presence of its absence, or the memory of its taste?

That taste is familiar even to a foetus, and it is the last desire of the dying. It is the tasteless secret of every flavour – cool, lukewarm, as hot as steam, or sharp and icy, on the tongue. Sometimes a quiver on the nerve endings of a tooth.

All living beings eat different kinds of food. But everything that lives slakes thirst with water.

From amoebae to birds to leopards to schoolchildren, from pigs to monkeys to flocks of birds to a woman waiting at the water's edge, all animals know the same tasteless taste. The taste of water.

Imagine thirst. And then you will know the taste of water. It's the memory of all memories.

Thirst – the arid absence of the humid. Negative moisture. A dryness of the body, and the soul.

No water. Not a drop. Not even a teardrop. Not a trickle. Not even a snowflake or a wisp of steam, or a bead of sweat, or even the sudden urge to pee, which is water asking for escape, a fugitive on the run from the flesh.

Before it is anything else, every life is a sac of water. An amniotic possibility. A spring. A source. A pool. A womb. A well. A step well.

That is why the word 'vessel' has two meanings: container, and boat.

Water on the inside, we are also, often, in water, on the outside.

Water bodies swimming in water bodies.

Coming up for air.

A well that stares at the sky is like an unblinking eye. A sliver of sky that bathes in its own reflection.

Sleeping beneath the earth, floating in the sky, descending like a gift, raining, raging like a storm, flowing, seeping, carving rock, oozing, freezing, boiling, scalding, crashing, drowning, rippling, quenching, quivering like a mirage of thirst on the horizon, water finds its level.

But in that time before life there was a time when there was none. Hot molten rock flowed, welled up like thirst, waiting for a rain of wet meteorites. Much of the earliest water on Earth rained down, as ice, mixed up with rocks from deep space.

There is a molecular gas cloud in the Orion Nebula 1,500 light years away that contains more than a million times the water held in the Earth's oceans.

There is another cosmic object, a quasar 12 billion light years away from us in the Lynx constellation, that contains 140 trillion times the water held in the Earth's oceans. This water was formed not long after the universe began, and it is still rippling. Still creating waves.

The Parable of the Step Well

There is a lot of water out there. But we all know what it means to be parched.

As things heat up, the oceans will rise, the rivers and ponds and wells will dry. There will be floods. There will be thirst.

What does a step well preserve in the marking of its steps? The memory of water?

What is the parable of the step well? You can climb down to the source in lean times. Or it can fill and come up to where you stand in the wet season. Water finds its level.

But how do you calibrate your thirst? How do you measure that thing that has no measure? ◆

Space as inventory, space as invention. Space begins with that model map in the old editions of the *Petit Larousse Illustré*, which used to represent something like 65 geographical terms in 60 sq. cm, miraculously brought together, deliberately abstract. Here is the desert, with its oasis, its wadi and its salt lake, here are the spring and the stream, the mountain torrent, the canal, the confluence, the river, the estuary, the river mouth and the delta, here is the sea with its islands, its archipelago, its islets, its reefs, its shoals, its rocks, its offshore bar, and here are the strait, the isthmus and the peninsula, the bright and the narrows, and the gulf and the bay, and the cape and the inlet, and the head, and the promontory, here are the lagoon and the cliff, here are the dunes, here are the beach, and the saltwater lakes, and the marshes, here is the lake, and here are the mountains, the peak, the glacier, the volcano, the spur, the slope, the col, the gorge, here are the plain and the plateau, and the hillside and the hill, here is the town and its anchorage, and its harbour and its lighthouse . . .

Georges Perec 🌢

(15)

To Study Water

is to Study Change

By Yasmine Hafez

When water changes, do we change too?

Lake Mariout in Egypt was once an important passage linking the River Nile to the Mediterranean, furnishing Alexandria with the economic and cultural legacy we know today. Over the past century Lake Mariout has drastically shrunk, fading at Alexandria's western borders due to years of land reclamation, urban expansion and industrial development encroaching upon its shores.

For years, I have researched the relationship between the Nile and the people who live alongside it, and what forms and reshapes this relationship over time. Initially my research focused on how Ethiopians perceive the Nile and how this affects their attitude to the Grand Ethiopian Renaissance Dam.

My research led me to lakes – ever-changing, delicate and fragile waterscapes that serve as sources of life and rich nutrients for the river. Lakes are economically significant for their fisheries, fertile lands and as sources of sustenance. But they can also endure slow deaths. Around this decay are stories of fear, resilience and loss – stories shaped by the demise of the lakes and their legacies.

When I began my fieldwork by Lake Mariout in 2022, a local fisherman asked me, 'You're studying this lake? This water? But why? It will soon disappear.' With Alexandria's sprawl pressing against the lake's edges, small rural communities struggle to remain fishing and farming when state interventions – such as roads and infrastructure projects – are cutting across the lake. The projects have exacerbated its depletion, distancing fishermen from the waters that have sustained their livelihoods for generations.

When I asked people about the lake's past they referred to it using the Arabic word *baraka*, which means divine blessings and abundance. Today the sense of *baraka* has faded, with resources in decline and fishermen forced to do their best to survive. This is the reality of experiencing a waterscape's slow death as it transforms from a vibrant ecosystem to a fragile one. As water disappears, people's memories, stories and legacies remain.

Lake Mariout is now isolated and shrinking. The waterscape itself – changing, shifting and slipping away – tells its own story. To live alongside it is to live with both memory and loss, with uncertainty and persistence. It is to live with change: water, memory and life. And so when we think of water, we must remember to make space for change and embrace the lack of certainty in its fluid essence. ◆

The Tubewell

By Anthony Acciavatti

In February 2013, with a fever and loss of nearly 8 kilos in nine days, I was admitted to Lenox Hill Hospital in New York City. On the same floor where Beyoncé Knowles supposedly gave birth to Blue Ivy, the doctors and nurses puzzled over my symptoms. I had stomach cramps, loss of appetite, dehydration, infrequent diarrhoea and fatigue.

The doctors asked me the usual question: had I recently travelled? I had just returned from a trip to India documenting the Kumbh Mela (Full Pitcher Festival), where approximately 80 million to 100 million people converge on a temporary tent city at the sacred confluence of the Ganges and Jamuna Rivers every twelve to thirteen years.

I hoped it wasn't giardia, which I had contracted once before. Giardia is a protozoan parasite that thrives in the small intestine. Transmitted through faeces by an infected person or animal, it is moderately resistant to chlorine and its low infectious dose makes it easy to contract from drinking water and recreational waters like lakes and ponds. In 2022, nearly 200 million people were infected with giardia worldwide and half a million reportedly died from it.[1]

Unfortunately, tests confirmed it was indeed giardia – but it was not the only parasite colonising my small intestine. Cryptosporidium, typically shortened to crypto, produces the same symptoms and was blooming inside me as well. Both parasites are routinely treated with Flagyl, an antibiotic that leaves a metallic aftertaste. After two rounds of Flagyl, the symptoms returned. Fortunately, a gastroenterologist requested a compound pharmacy in New Jersey to make a special pill to kill off the parasites cohabitating in my body. It worked.

What does my hospitalisation after visiting a major festival in India along the banks of the Ganges and Jamuna have to do with a book on thirst? What might these parasites tell us about groundwater and public health?

I was in India wrapping up fieldwork for my first book, *Ganges Water Machine: Designing New India's Ancient River*, which investigates how this basin came to be the most populous and hyper-engineered landscape on the planet. A long-standing concern across the Ganges, and increasingly across much of the world, is the contamination of lakes and rivers by faecal matter from humans and non-humans alike. This is a particular anxiety at the temporary tent city constructed for the Kumbh Mela, which relies on what has become an everyday technology to transform groundwater into infrastructure: the tubewell.

Mechanised technologies for drawing up groundwater is the second largest distributed mass of freshwater on the planet. Tubewells are used in places where municipal water supply is non-existent, unreliable or polluted. Composed of perforated steel or plastic pipes that are bored between 1.5 and 300 metres below the ground to extract water from an aquifer, tubewells are typically driven by an electric or oil-powered engine.

In 2013, numerous tubewells were sunk for the Kumbh Mela to ensure potable water for the millions of visitors and to reduce the chances of waterborne illnesses like giardia and cryptosporidium.[2] And they are routinely used at this site every year during the annual festival of Magh Mela, which also takes place in January and February. While tubewells are a temporary public-health technology for these festivals, they, along with hand pumps, are permanent fixtures across India's farms and cities.

India extracts 259km^3 of groundwater every year. This is not only more than China and the United States combined, but it accounts for almost one-third of the global total. Most of India's groundwater is used to irrigate crops, while only 9 per cent is for drinking; in Indonesia that figure is 93 per cent. Hydrologists estimate that nearly half the global population drinks groundwater daily and that over half of the world's irrigated crops rely on it.[3] Growth in agricultural production and population, as well as the impact of climate change, will undoubtedly increase these percentages.[4] In short, groundwater extraction technologies are as significant to farms and cities as the elevator was to the early-twentieth-century American metropolis.

Our thirst for groundwater has shifted the axis of the Earth – between 1993 and 2010, according to scientists' estimations, by at least 80 centimetres.[5]

Despite scientists' ability to calculate the annual 4-centimetre drift induced by tubewells and hand pumps, we have no way of knowing how many of these devices operate today. In most parts of the world, groundwater extraction is unregulated. This makes aquifers particularly susceptible to contamination by microbial parasites found in wastewater, as well as agricultural fertilisers and industrial effluents.[6] What is more, the vast majority of tubewells are privately owned and operated. Millions of tubewells and hand pumps have allowed billions of people to thrive, but they also threaten to bleed portions of the Earth dry.

The effects of this are most clearly visible in the form of sinkholes and kilometre-long fissures. When the number of tubewells multiplies in a given area and groundwater is pumped out faster than it can be replenished, subsidence often occurs; the water-table level drops and the soil compacts. Soil compaction is also compounded by the weight of buildings, which are not designed to move more than a few centimetres, let alone sink by several metres. The effects of subsidence are visible across cities as diverse as Addis Ababa, New Delhi, Jakarta, Phoenix and Mexico City.

Subsidence, like a cyclone or a tornado, cuts across private and public space indiscriminately. Roads collapse, buildings sink, and land separates due to the extraction of groundwater. It raises questions about insurance, responsibility, and ultimately who pays for repairs. In much the same way that cars and factories impact air quality and public health, private extraction of groundwater has public consequences. Yet we do not have conventions for monitoring the subsurface of the earth in the way that we have for particulate matter in the air.

How might we visualise the ways in which we extract water from the subsurface in order to terraform the surface?

Imagine we were to make an X-ray of the New Delhi Metropolitan Region so that we could see the hundreds of thousands of tubewells in relationship to the settlement patterns of over 30 million people – not only in terms of the form of the city, but also the layers of the city such as major roads, the metro system and surface water bodies like rivers and lakes. If we made such an X-ray, not only of New Delhi but also of Jakarta and the Phoenix-Tucson megaregion, we would see how tubewells tap into a vast commons that rivals the air we breathe.

It is imperative that we see and draw groundwater differently, as a collective resource

The Tubewell

that undergirds the shaping of cities in the twenty-first century. Tying the shape and life of cities to groundwater is the only way to envision a future that does not rely on the private pump. Although communities have come together to reverse subsidence and replenish aquifers in places like Dwarka, just south of New Delhi, subsidence is not the only concern we should collectively share.[7] With more and more tubewells and a lack of municipal wastewater management, we run the risk of protozoan parasites like giardia and cryptosporidium entering the water supply along with E. coli and salmonella, to name just a few. Making drugs like Flagyl readily available without a prescription certainly helps millions of people who lack access to medical testing; however, we are seeing the limits of this approach. For at least a decade there has been an increase in cases of drug-resistant giardia.[8] Our thirst for groundwater and dependence on the private pump are at odds with one another. When millions of private tubewell owners are extracting groundwater with little to no regulation, contamination from polluted surface water bodies will only proliferate. ◆

Notes

1 S. T. Hajare, Y. Chekol and N. M. Chauhan, 'Assessment of prevalence of Giardia lamblia infection and its associated factors among government elementary school children from Sidama zone, SNNPR, Ethiopia', *PLoS One*, 17(3) (15 March 2022): e0264812, doi: 10.1371/journal.pone.0264812. PMID: 35290402; PMCID: PMC8923448.

2 Concerns about an outbreak of an epidemic disease like cholera during the festival date back to the nineteenth century. See Kama MacLean, *Pilgrimage and Power: The Kumbh Mela in Allahabad, 1765–1954* (New York: Oxford University Press, 2008), pp. 5, 12, 32.

3 Jean Margat and Jac van der Gun, *Groundwater Around the World: A Geographic Synopsis* (Boca Raton: CRC Press, 2013), p. xix.

4 Richard G. Taylor, Bridget Scanlon, Petra Döll et al., 'Ground water and climate change', *Nature Climate Change*, 3(4)(2013), pp. 322–9.

5 See Ki-Weon Seo, Dongryeol Ryu, Jooyoung Eom et al., 'Drift of Earth's pole confirms groundwater depletion as a significant contributor to global sea level rise 1993–2010', *Geophysical Research Letters*, 50(12) (June 2023), https://doi.org/10.1029/2023GL103509; Warren Cornwall, 'Humanity's groundwater pumping has altered Earth's tilt', *Science*, 16 June 2023, doi: 10.1126/science.adj2812.

6 M. E. Daniels, W. A. Smith and M. W. Jenkins, 'Estimating cryptosporidium and giardia disease burdens for children drinking untreated groundwater in a rural population in India', *PLoS Neglected Tropical Diseases*, 12(1) (29 January 2018): e0006231, doi: 10.1371/journal.pntd.0006231. PMID: 29377884; PMCID: PMC5805363; A. Rahman, N. C. Mondal and K. K. Tiwari, 'Anthropogenic nitrate in groundwater and its health risks in the view of background concentration in a semi-arid area of Rajasthan, India', *Scientific Reports*, 11, (9279) (2021), https://doi.org/10.1038/s41598-021-88600-1; A. Quddoos, K. Muhmood, I. Naz et al., 'Geospatial insights into groundwater contamination from urban and industrial effluents in Faisalabad', *Discover Water*, 4(50) (2024), https://doi.org/10.1007/s43832-024-00110-z.

7 S. Garg, M. Motagh, J. Indu et al., 'Tracking hidden crisis in India's capital from space: Implications of unsustainable groundwater use', *Scientific Reports* 12, (651) (2022), https://doi.org/10.1038/s41598-021-04193-9.

8 E. R. Carter, L. E. Nabarro, L. Hedley and P. L. Chiodini, 'Nitroimidazole-refractory giardiasis: A growing problem requiring rational solutions', *Clinical Microbiology and Infection*, 24(1) (2018), pp. 37–42, ISSN 1198-743X, https://doi.org/10.1016/j.cmi.2017.05.028; S. Krakovka, U. Ribacke, Y. Miyamoto et al., 'Characterization of metronidazole-resistant Giardia intestinalis lines by comparative transcriptomics and proteomics', *Frontiers in Microbiology*, 13 (10 February 2022);13:834008, doi: 10.3389/fmicb.2022.834008. PMID: 35222342; PMCID: PMC8866875.

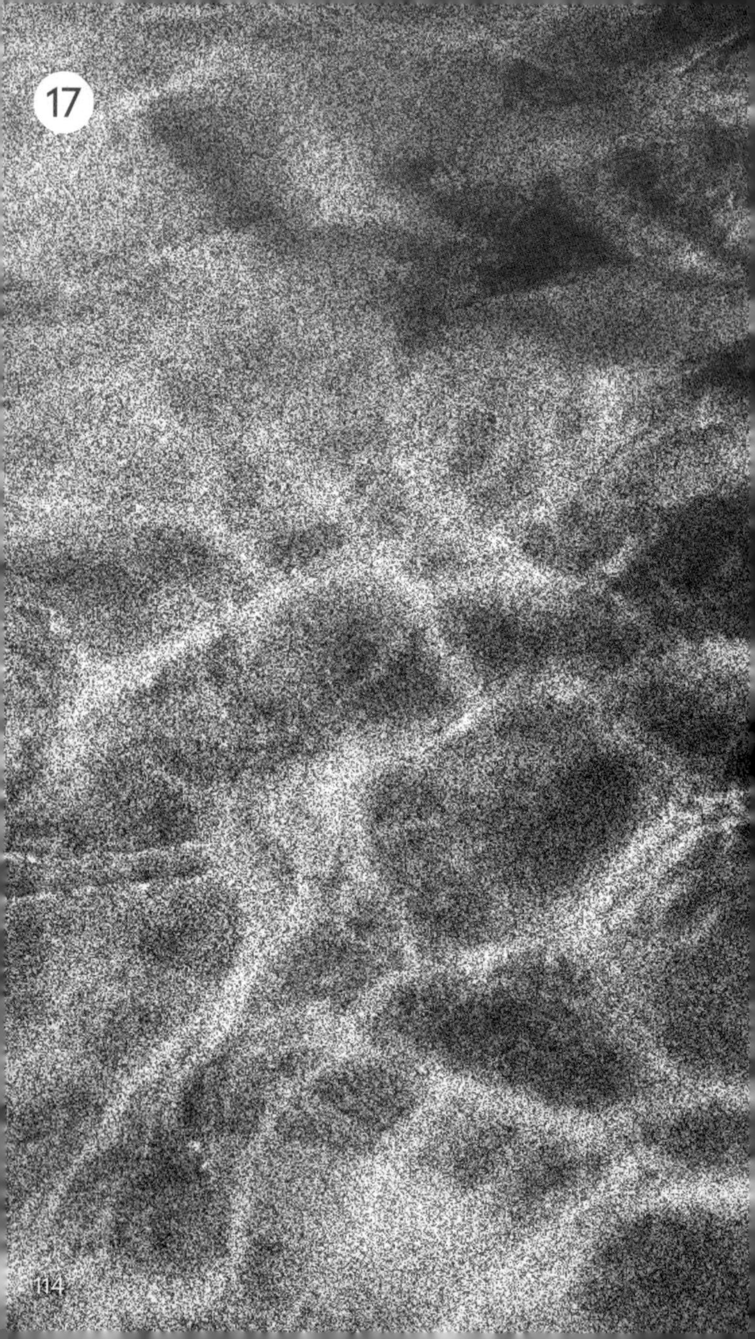

17

Murmur

Nightmares

By Karan Shrestha

to the

Stream

you say mother –

>   water
>   flowing drowns the dread.

But it rains snakes again.

Snakes swallowing helicopters and stars whole,
>   while cascading down. Slender wet bodies
kissing tiled roofs, constricting cement walls to powder.
>   My escape, always narrow,
>   is a pattern for flesh to stay snug around bones.
>   The surprise is how I am balancing
a clay pot on the run,
with a stray dog jaw locked into my right thigh.
>   My blood red
>   on his fur umber
that coats four upturned bottles dangling empty.
I tug at the stray dog's tail and he unzips.
>   There are glass-necks for limbs that clank
sharper than Earth rumbles.
>   I dodge the slide of lands, clasping
>   the clay pot where
>   grandmother finally sprouts. A citron plant –
>   she chooses
>   to be an ancestor. Sour will be her water.
We drink what we can of desire. We crisscross
>   over government pipelines that keep us thirsty.

I open the tap
– speak to water. Speak-water is a one-way channel
    billowing words to shape droplets birthing
calloused moons.

        Dream-snakes are guides. Luck bringers,
you say mother.
          How much can water keep?

Every twelve years the river returns, you say mother.
          What holds water?

Murmur nightmares to the stream,
you say mother –
        water
        flowing drowns the dead.

And the reservoirs are all afloat.

Karan Shrestha

By now we are familiar with being
            robbed of indigo skies.
Like tattered blankets, ripped and holed,
                and torched by flaming crimson and
neon green eyes
        of giant *Karkotak* – the slithering vehicle
fetching fauna of the valley,
                        shrinking to slide effortless into
abandoned burrows.
Karkotak migrates to leave behind cities in fumes.
Outside the last preserved bureau,
                pointed nose men with topis,
            unbuttoned suits, caw
                in circles. Pants down, piss in trickles.
They stand around a pool of lotuses,
            regurgitate incantations to quell the quakes.
                        Stale breath makes the sacred
flowers wilt. So the men
        morph to crows with clipped wings instead.
I watch them skip through the ruins.
        Why did they forget the mountains
were carved by ice?
            No spell now could avert the deluge.
Taudaha tamed to a pond, swells to recover
Kathmandu,
            once a lake. Future sea. Gulping debris.

18

Divining

By Gaylene Gould
and Calthorpe Community Participants

Mary

'

Every summer as children, our
playing space was the river in
the Carnic Alps. It ran behind
my grandmother's house. Jade,
blue, green, full of stones. Every
time I return to visit her, she
teaches me about the constancy
of change; she's never the same.
She's an artist, sculpting stones
and her own bed.

'

Calthorpe Community Participant

I've been without water or heating for three days now and it's the dead of winter. My whole block is without water, in fact the entire street. Thames Water are desperately trying to fix a slew of problems – from a broken valve in Brixton to air bubbles in the pipes that had turned our water suspiciously cloudy – and have now staunched the flow entirely. My neighbours and I are filling buckets with water from an outside standpipe to flush our toilets and scouring shop shelves for bottles of water to drink. We are huddling in our single rooms around electric heaters and I'm grimy from not having washed properly for a few days. I'm extraordinarily lucky that throughout my lifetime water has been available at the turn of a tap and I'm confident that this situation will be short-lived – so why am I gripped by a primal sense of terror?

Thames Water, and other water companies, have been trending of late. Stories of failing infrastructure, pollution pumped into rivers, a company on the brink of financial collapse have kept London's primary water company in the headlines. There are many roots to this cataclysmic situation, including higher levels of

Divining Mary

rainfall flooding the system. However, private ownership is clearly not helping. The profits that should be spent on necessary upgrades are being split with shareholders. I realise that my growing sense of alarm is not for this situation alone but in anticipation of a future of forced thirst.

> 'The River Lea that fed the canals near the marshes was the closest body of water. It overflowed often and taught me about the unpredictable nature of water but I loved watching the sunset reflected in it.'
> *Calthorpe Community Participant*

London's landscape, its rises and falls, is shaped by the memory of its rivers, many of which flow unseen. The River Fleet is one such. Since Roman times, the Fleet has run from the north of the city, now Hampstead, issuing into the Thames at Blackfriars. Today the Fleet continues to flow down this route as part of the sewer system and beneath generations of concrete. Outside a pub on Clerkenwell Road you can still see, hear and feel the power of the rushing river through a grate in the pavement. This river was once known as the River of Wells because it fed healing water wells along its banks. The healing water, or chalybeate, came about by the river water mingling with the dense, mineral-rich London clay.

In the 1600s there was a well fed by these healing waters, known as Black Mary's Well, situated in a marshy rural area, now modern-day King's Cross. The eponymous Mary is a legend. There are many conflicting accounts of the mysterious woman whose name was embedded into the landscape long after she disappeared. The best description can be found in

Thomas Cromwell's 1828 book *History and Description of the Parish of Clerkenwell:*

> Beneath the front garden of a house in Spring Place, and extending under the foot-pavement almost to the turnpike-gate called the Pantheon Gate, lies the capacious receptacle of a Mineral Spring, which in former times was in considerable repute, both as a chalybeate, and for its supposed efficacy in the cure of sore eyes. [...] About one hundred and forty years since, it was the occasion of giving the elegant epithet of Black Mary's Hole to a small old house on this, and a few others on the opposite side of the road, from the following circumstance. In the single house eastward of the road [...] immediately contiguous to the fountain, lived a black woman named Mary Woolaston, who rented this spring, or conduit, of Robert Harvey, Esq., and made a living by the sale of its water to the citizens.

There are sightings of Mary Woolaston, or Black Mary, in other historical records. Some question her existence. The trail quickly runs cold. There are descriptions of the healing wells – but most from 100 years or so after Mary's era, when the wells had become a commercialised part of London's social fabric.

The wells seemed to have served the mixed function of a place for healing and a lively social hangout. They were accessible places for everyone – rich and poor – and were often lively and rambunctious. Cromwell goes on to note that by the time Mary died, Black Mary's Hole had 'obtained a character for licentiousness'. Later the area became a popular gay cruising spot.

Given that the well-keepers would have held space for a broad community of people with physical, mental and social needs, the work must have been fascinating. Yet there is little on record about their lives.

We do know about the men who profited from the waters, however. Their lives are well documented. Men like Baynes who, after Mary died (around 1687, one record states), redirected the spring to his new spa development called the Cold Bath. Then there's Hughes, who bought a neighbouring house, Bagnigge Wells, in 1757 and, once he discovered there was a mineral spring on the premises, promptly opened a pleasure garden. Pleasure gardens were a commercial trend which turned the wells into circus-like attractions – places to dress up, be seen and, often outrageously, entertained. Like most entertainment centres, the gardens soon lost their glow and fell into disrepute, which forced the well water underground for the last time.

The most prestigious 'man of water' of the time was undoubtedly Hugh Myddelton. In 1613, he opened London's first water station, right behind Mary's spring at around the time she was alive. The New River Head miraculously pumped clean water into London via a conduit from Hertfordshire, an engineering feat spearheaded thanks to Myddelton's powerful connections. These included his Lord Mayor brother, Thomas Myddelton, and King James I. The king agreed to cover half the costs of the enterprise in exchange for half the profits. In fact, to fund the New River Head, the Myddeltons created one of the first joint-stock utility companies, basically a company led by shareholders, which turned water into a capitalist venture for the first time. The trio's entrepreneurial zeal moved from water to land and people. As the New River Head was being constructed, James I was busy establishing the Plantation of Ulster, stealing land from the Irish to hand over to the English. Thomas Myddelton, meanwhile, was founding the East India Company, which went on to rule large areas of the Indian subcontinent. This was the Age of Discovery; white men were beginning to realise that they could control water, land and people for profit. While it's nigh impossible to find information about the independent working well-keepers like Mary Woolaston, there are multiple streets, squares and statues in honour of Hugh Myddelton. And the New River Head? Well, that became the Metropolitan Water Board, which in turn became Thames Water.

> 'Credit Valley River – the same river can invoke different memories depending on the area of land it is connected to.'
> *Calthorpe Community Participant*

I'd first read about Mary Woolaston in my friend Nana Ocran's 2003 guide *Experience Black London*. Today such a guide would be researched and shared online. But this was a few years before the internet had become a global archive. It was also many years before historian David Olusoga would appear on primetime TV popularising Black history. So Nana's opening chapter, which traced Black presences in Britain from Roman times on, was a much more singular effort to compile than it would be today. I have a mental image of Nana flipping microfiche after microfiche into a machine at the London Metropolitan Archives.

Now, when I read the small paragraph that captivated me back then, it's hard to fathom why this section stuck. '"Black Mary's Hole" was a lonely spot on the road from London to Hampstead near Cold

Bath Fields. The place was named after "a Blackamoor woman" called Mary, who lived by the side of the road in a small hut built with stones.'

There were mentions of the extraordinary freedom-fighter Olaudah Equiano and Henry VIII's court musician John Blanke, but it was the image of this mysterious Black woman, living freely by the side of a road named after her, that seeped into me. Over the next twenty years, I continued to search for traces of Mary. The more information I unearthed, the more my obsessions were compounded. There were healing water wells in London and one kept by a Black woman? Who was she? Where did she come from? What drew her to this place and work? How did she dispense the water? How did she feel while doing it? How did she fare out there alone as a sole Black woman? Was she the sole Black woman? After she retreated into her stone dwelling at the end of a working day, what did she dream about?

I am a Black woman. I am an artist-facilitator who has spent a great deal of time creating healing spaces for others. It's a strange compulsion to want to help others to recuperate. My Caribbean mother was a nurse, as were many Black women of her generation. Maybe I inherited a sense of service from her, as she might have from her housemaid grandmother. Caring work is rarely paid well or publicly valued. It can lead to deep feelings of connection with others but can also be extraordinarily depleting. And given the particular way in which our society has constructed race and gender roles, the Mammy, the carer, the wet-nurse are ones that Black women can more easily occupy. These jobs suggest selflessness and, in that, a disappearance of self. Black women are often treated as *verb*, as function, rather than with the tenderness of a complex being with precious interiorities. These tensions increasingly trouble and inspire me and I wonder if they did Mary too. In other words, the more I researched Mary, the more I began to research myself.

'Walaga – flow.'
Calthorpe Community Participant

To locate water by divination or dowsing is to walk across solid ground brandishing a forked stick or pendulum. If there is hidden water beneath, then the implement will shudder and tremble. Some say water dowsing works, because of how water energetically affects our bodies, a psychological effect known as the ideomotor phenomenon. Some believe there is a direct exchange of energy between water and the physical body, maybe a co-mingling of outer and inner waters.

Historians will most likely disagree, but I think historical research is akin to water dowsing. Research is an act of divination, led as much by feeling as by evidence. I think of Philippa Jayne Langley, as she searched for Richard III in my home town of Leicester, walking into a car park and feeling an intense bodily sensation. Without any physical proof, she knew immediately where the missing monarch's bones lay and years later would go on to prove it. The search for a king is one thing, but seeking out early Black or working-class presences is another altogether. Here we have to rely heavily on sensing and intuition due to the lack of care in evidencing a subject like Mary. Theorist Saidiya Hartman describes the intangible ways in which the lives of the people uncared for by history are forced to be recalled. She uses the term *critical fabulation* to describe the need to equally research and imaginatively conjure up these shadowy characters.

As we began to shape our own critical fabulation project, reimagining Mary and exploring her contemporary resonances, we started to pay attention to the many serendipitous messages that emerged. When they did, we would say, 'That's our Mary.' During a rain-drenched week, the sun would suddenly shine brilliantly on the day we held a public event. 'That's our Mary,' we'd say. Mary's spirit became our most important collaborator.

The arrival of Emanuela Aru was a major 'That's our Mary' moment. Emanuela, a lapsed history student, in tender recovery from the brutal pandemic years, had popped in for a tai chi class at Calthorpe Community Garden, where we were based, and saw a flyer for our meeting. Emanuela had already begun to do her own research into Black Mary and the healing wells by then and so was astonished by the sudden appearance of our project. She is tenacious, curious and passionate, and so it made complete sense to obey Mary and engage Emanuela as the project's researcher. Over the next year, Emanuela voraciously 'divined' the archives while building relationships with researchers at Historic England, attending local meetings and uncovering all she could on Mary. It was Emanuela with whom I would go on to co-write a Healing Tour, which offered participants a chance to search for Mary following Emanuela's 'clues' while sharing their own stories of water healing. Emanuela is now a trained tour guide with her own business and most of the facts in this essay are thanks to her research. She says that it was searching for Mary that led her back to herself in the same way that Mary was leading me back to myself. Four hundred years after Mary disappeared, here she is, healing us still.

It was a bright and brisk November day in the wake of the pandemic when we went in search of Black Mary's Hole. Recovering the memory of public healing spaces now felt like an urgent and vital endeavour. The world was in need and it was time to find Mary. We had scoured the internet for clues as to where the well might be and Zaynab, my friend and project producer, had sketched out a potential route that might lead us there.

We started at Farringdon Station, a surprisingly bustling area in the heart of old London. Pret A Manger eateries were jammed up against 400-year-old pubs, leaving us disoriented in time. This is the inner city, parched, traffic-clogged, with some of the worst air pollution in London and with absolutely no evidence of water, healing or otherwise. But our route plan suggested that there was a place called Wells Court on Farringdon Lane, where remnants of the old Clerks Well remained.

We eventually found the ancient stone casing housing the deep black hole that leads down and back into time. Unbeknownst to the striding lunchtime crowd, 900 years ago parish clerks were performing their mystery plays in this very spot. Now you need to cup your hands against the glass of a nondescript 1980s office block to get a glimpse of the quietly heroic structure, an aged stone hole leading us deep beneath the paving.

Farringdon Road became King's Cross Road – a main north–south artery, and one of the most noise-polluted. At times we had to raise our voices to be heard above the thundering lorries as we walked beside each other. Even though I've lived in London for over thirty years I've never got used to the constant racket of the city, so when Zaynab led us off the main road on to the surprisingly quiet Conduit Street I was relieved.

There were clues. The shift in quiet on Conduit Street, the soft bend of the road, like a river might make. We spoke about the way in which the memory of water marks our contemporary landscape in street names like Conduit and look, there! Wells Street.

I didn't know then that the Fleet was running beneath our feet, and maybe it was this that was causing me to tremble. 'She was close to here,' I said, surprising myself. 'I can feel it.'

We allowed our instincts to lead us on, rounding the corner on to Calthorpe Street, a tidy strip of Georgian houses, an inexplicable sense of excitement growing. At the end of the street the quiet erupted into another busy road, this time Gray's Inn, three lanes of traffic hurrying on. As we turned down the street our attention was pulled to the left, towards an inconceivable patch of green. As we walked through the impressive red and black wooden entrance emblazoned with the name Calthorpe Community Garden, the tingling began to settle into another feeling which could only be described as knowing.

Over the next few years, we would go on to partner with the Calthorpe Community to reflect on Mary's legacy through a fluid and responsive participatory art project. At the centre is the creation of a healing garden led by the community and a new memorial commemorating Mary, by artist Marcia Bennett-Male. We would learn just how this improbable space, in this overbuilt area among some of the most expensive real estate in the country, had been fiercely guarded and cultivated by the local community despite encroaching commercial concerns. We would learn from the founder, Annika Miller Jones, and long-term community worker Mila Campoy, as well as other staff and volunteers, about the energy, resilience and kindness that it takes to keep a modern-day healing well.

Calthorpe draws people from elsewhere – the recently arrived and those looking for a home in a city that increasingly offers few free welcoming spaces. The shifting community reflects the global shifts in the world. Afghani women living in local hostels would regularly gather to cook delicious food together in the large open kitchen. More recently, the Hong Kong and Ukrainian communities have taken up residence and the Latin American elders are the soul of the place. Calthorpe – with its busy food-growing garden and benches where teenagers peacefully eat their lunch and children chase each other in and out of nooks and crannies while their parents gather round the large dining table to share meals; and for those not well enough to join in, food parcels are sent. This is Mary's land, and it is evident that the healing well waters still run in the ground at Calthorpe. Now, though, Mary is no longer alone and no longer forgotten. She has birthed many, many well-keepers.

All this we would discover later.

That first day, we stopped on the wooden bridge that links this timeless garden to the outside. Running beneath us, through a rough and wild sunken garden shielded by tall, cool trees, was a small stream that seemed like an impossibility in this time and in the place that London has become. And yet here she was. ◆

water

sign
woman

By Lucille Clifton

the woman who feels everything
sits in her new house
waiting for someone to come
who knows how to carry water
without spilling, who knows
why the desert is sprinkled
with salt, why tomorrow
is such a long and ominous word.

they say to the feel things woman
that little she dreams is possible,
that there is only so much
joy to go around, only so much
water. there are no questions
for this, no arguments. she has

to forget to remember the edge
of the sea, they say, to forget
how to swim to the edge, she has
to forget how to feel. the woman
who feels everything sits in her
new house retaining the secret
the desert knew when it walked
up from the ocean, the desert,

so beautiful in her eyes;
water will come again
if you can wait for it.
she feels what the desert feels.
she waits.

Thirst

By Lucy Jones

The sun is low in the sky. The sky is blue. The road behind me whirrs with cars. The people are going somewhere: to work, perhaps driving home after the school run. In seconds, I move from the car park right on the busy A road to the river. A lot of rain has fallen recently, and the river breathes on to the path. A splosh. A vole? At the gate, rubbish is piled on one side: Lucozade, Powerade, Rio, Strongbow, a black Nike trainer, a vape oil, Appletiser, Stella, Coke. Through the gate and the path is edged with willow, alder, willowherb, angelica, nettles, brambles, reeds.

As soon as I see the water I feel instantly relaxed, instantly better. It is flowing fast today.

I thought rivers were for everyone, or at least that you couldn't own a river. After moving to this part of the world, a county in southern England rich in woodlands and chalk streams, I became attached to a few, particularly in times of mind-sickness when only the cold river

Thirst

could shake my anhedonia. But soon we were scolded by anglers and I learned that many of the local rivers were in fact privately owned, and that we were trespassing and might disturb the trout that men who lived as far as New York paid a licence each year to fish. I needed to find another river to spend time with, a place I could be without worry about trespassing.

This river is the one that we found. It is a place intentionally set aside for recreation and biodiversity. Officially a SANG – a Suitable Alternative Natural Greenspace – it was initially provided, and is managed, to keep people and dogs away from more vulnerable sites which might have ground-nesting birds, say, or other protected species.

It smells salty, mineral. Fresh. The mud squelches and the water seeps into my no-longer-waterproof shoes.

As I walk further from the road I start to hear more birds and my eyes adjust and I see insects and spiderwebs, crossing the path. I can feel the spiderwebs on my face as I walk. Soft and strong and gentle.

There are small desire paths where dogs or people venture into the river between the reeds. In the summer, children and teenagers hang out in the gaps: playing, chortling, flirt-ing, hooting. No one else is here today, and so my attention is purely on the river and its wider ecosystem. The water reflects the light of the sun: glimmering, flashing, like static. Nettles stand to attention like sentries, slightly blackened by a frost, but alert. I see the yellow blotches of plasmodium on a fence post.

I follow the river as it curves and bloats. I love its wiggle, and find out when I speak with the ranger that the curve has been put back in by his team, to rewild the area and make it more habitable for species. Like most rivers in this country, agricultural run-off has damaged ecologies, but the local Wildlife Trust works to patch it up, with gravel and other clever conservation strategies. In fact, this river has 'poor' ecological status, because of sewage, agricultural land use, other pollutants and the bank-eroding presence of signal crayfish.

To my eyes, the river is beautiful. Today, spiders skate on the surface; goldcrest and long-tailed tits strip the trees above. In the summer it is filled with mayflies, dragonflies, damselflies and the smell of watermint.

A dog barks. A dog whistle. A woman calls for her dog. Above, an airplane. A train. I can see through the train windows, like I can see through most of the tree's branches, back to the blue. Above, eight geese fly.

I sit on the damp bank, the water of my body magnetised to the water of the river. The yellow leaves take the light like scraps of coloured cellophane or stained glass.

The moving water sounds so good. I wonder how you would write it phonetically.

Lblblblblblblblblblblblblb

Rblrblrblrblrblr

Blblblblblblblblblblblblblbl

Bwub bwub bwub

The whole of the world is this river.

I could sit and watch the patterns all day.

Eddies and swirls,

circles and lines,

ridges and stripes.

It is in my brain, licking hot sad areas, dissolving knots of strain and worry. Nurturing; soothing.

The river is a mother.

Part of it looks viscous or gelatinous where reeds snag. Logs and plants create obstacles which create other sounds and patterns. When a bird takes a twig or drops a twig, the patterns change. Parts are ribboned, striated, corrugated. I am very still. Around a root the water becomes white, frothy, bubbly. A theatre of ASMR in front of me. I think this is 'soft fascination', an aspect of Attention Restoration Theory which explains why being in the natural world is restorative and soothing for mental fatigue.

The bang of farm machinery. A siren. A red woodpecker zips in and out. A lorry. I strain to tune into the burble of the river. I wish I could hear the river more, without the background noise, but I remember that natural sounds are powerful, perhaps even if we can barely hear them: one study showed that even under anaesthetic, *even when unconscious*, people produce fewer stress biomarkers if they're played natural sounds, such as water flowing.

I walk back a bit to the area where we take the children on very hot days. Living motes – lice, perhaps – skip across the surface. As does a fly, and spiders. I fill a small pot with water, to look at under the microscope. I want to see what else is living in the river, unseen to my human eye. Later I will see another world: microscopic animals, rotifers, amoebas: creatures with spiky membranes, others that spit out tiny seeds as they spin at speed, sacs of amorphous flubber which move with determination.

I return to my car, river-soothed. No swim today, but I've received the gifts of the river: it's cooled the heat of a busy mind.

'Slake' is from the Old English *sleacian*, meaning to 'become slack or remiss; relax an effort'. It's related to the words 'slack' and 'lax', to make slack or loosen a rope of bridle, and later, to quench or extinguish a fire. *Ācwincan* is the Old English 'to quench', meaning to extinguish, to vanish, to be eclipsed.

Even on the banks, the river quenches the heat of the living, loosens the strains of consciousness, slakes the thirst for release. ◆

تشنگان گر آب جویند از جهان
آب جوید هم به عالَم تشنگان

مولانا جلال‌الدین رومی
مثنوی ۱, ۱۷۴۱

*For if the thirsty search for water, then
the water is, too, seeking the thirsty*

Jalaluddin Rumi

Masnavi, Book 1: 1741

About the Authors

Robert Macfarlane

is internationally renowned for his writing on nature, people and place. His bestselling books include *Is a River Alive?, Underland, Landmarks, The Old Ways, The Wild Places* and *Mountains of the Mind*, as well as a book-length prose-poem, *Ness*. His work has been translated into more than thirty languages, won prizes around the world and been widely adapted for film, music, theatre, radio and dance. He has also written operas, plays and films including *River* and *Mountain*, both narrated by Willem Dafoe. He has collaborated closely with artists including Olafur Eliasson and Stanley Donwood, and with the artist Jackie Morris he co-created the internationally bestselling books of nature-poetry and art, *The Lost Words* and *The Lost Spells*. As a lyricist and performer, he has written albums and songs with musicians including Cosmo Sheldrake, Karine Polwart and Johnny Flynn, with whom he has released two albums, *Lost in the Cedar Wood* (2021) and *The Moon Also Rises* (2023). In 2017 the American Academy of Arts and Letters awarded him the E. M. Forster Prize for Literature, and in 2022 in Toronto he was the inaugural winner of the Weston International Award for a body of work in the field of non-fiction. He is a Fellow of Emmanuel College, Cambridge.

Natalie Diaz

is a Mojave/Akimel O'odham poet, language activist, educator and former professional basketball player. Her poetry collections include *Postcolonial Love Poem*, winner of the Pulitzer Prize for Poetry; and *When My Brother Was an Aztec*, winner of an American Book Award. She is an enrolled member of the Gila Indian Community and lives in Mohave Valley, Arizona, where she has directed a Mojave-language revitalisation programme, and currently teaches in the Creative Writing MFA programme at Arizona State University.

Rebecca Solnit

is a writer, historian and activist, and the author of more than twenty books. They include *Orwell's Roses; Recollections of My Non-existence; Hope in the Dark; Men Explain Things to Me; A Paradise Built in Hell: The Extraordinary Communities That Arise in Disaster* and *A Field Guide to Getting Lost*. A product of the California public-education system from kindergarten to graduate school, she writes regularly for the *Guardian*, serves on the board of the climate group Oil Change International, and recently launched the climate project Not Too Late (nottoolateclimate.com).

Elif Shafak

is an award-winning British-Turkish novelist whose work has been translated into fifty-five languages. Her latest novel, *There Are Rivers in the Sky*, was a top-five *Sunday Times* bestseller. Her previous novel, *The Island of Missing Trees*, was a top-ten *Sunday Times* bestseller, a Reese Witherspoon Book Club pick and was shortlisted for the Costa Novel Award and the Women's Prize for Fiction. *10 Minutes 38 Seconds in This Strange World* was shortlisted for the Booker Prize and the RSL Ondaatje Prize, longlisted for the Dublin Literary Award, and chosen as Blackwell's Book of the Year. She is a Vice-President of the Royal Society of Literature. Shafak was awarded the Halldór Laxness International Literary Prize for her contribution to 'the renewal of the art of storytelling'.

Vandana Shiva

is a world-renowned environmental thinker, activist, feminist, writer and science policy advocate. Initially trained as a physicist, she later shifted to inter-disciplinary research and in 1982 founded the Research Foundation for Science, Technology and Ecology, an independent research institute that addresses the most significant ecological problems of our times. She is the recipient of many awards, including the Right Livelihood Award (the 'Alternative Nobel Prize') and the Sydney Peace Prize, serves on the board of the International Forum on Globalization and is a member of the executive committee of the World Future Council. She is the author of over twenty books.

Emmanuel Vaughan-Lee

is an Emmy- and Peabody Award-nominated filmmaker and a Sufi teacher. His films include *Earthrise, Sanctuaries of Silence, The Atomic Tree, Counter Mapping, Marie's Dictionary* and *Elemental*. His work has been screened at New York Film Festival, Tribeca Film Festival, SXSW and Hot Docs, exhibited at the Smithsonian Museum and featured on PBS *POV*, National Geographic, and *New York Times* Op-Docs. He is the founder and executive editor of *Emergence Magazine*.

Lora Aziz

is a British-Egyptian interdisciplinary artist, nature storyteller and creative producer whose work bridges ethnobotany, ecoliteracy and climate justice. Through her practices, she explores the connections between intergenerational knowledge, traditional practices and the natural world, crafting immersive projects that blend storytelling, research and art. Lora's work often involves foraged plant matter, transforming it into handmade inks, papers and dyes to tell regenerative stories of place and resilience. She is also a grower on her farm in Suffolk, where she cultivates a deep relationship with the land. Recent projects include *Homeland is Presence*, a publication celebrating cultural heritage through plants, and a COP27 creative climate commission.

Ocean Vuong

is the author of the critically acclaimed poetry collections *Night Sky with Exit Wounds* and *Time Is a Mother*, as well as the *New York Times* bestselling novel *On Earth We're Briefly Gorgeous*. A recipient of the American Book Award and the MacArthur 'Genius Grant', he has also worked as a line cook, tobacco-harvester, nursing-home volunteer and fast-food server, the latter becoming inspiration for his latest novel, *The Emperor of Gladness*. Born in Saigon, Vietnam, he currently splits his time between Northampton, Massachusetts, and New York City.

Robin Wall Kimmerer

is a mother, scientist, decorated professor and enrolled member of the Citizen Potawatomi Nation. She lives in Syracuse, New York, where she is a SUNY Distinguished Teaching Professor of Environmental Biology, and the founder and director of the Center for Native Peoples and the Environment.

Jessica J. Lee

is a British-Canadian-Taiwanese author, environmental historian and winner of the Hilary Weston Writers' Trust Prize for Nonfiction, the Boardman Tasker Award for Mountain Literature, a Banff Mountain Book Award and the RBC Taylor Prize Emerging Writer Award. She is the author of three books of nature writing, *Turning, Two Trees Make a Forest* and *Dispersals*, the children's book *A Garden Called Home*, and co-editor of the essay collection *Dog Hearted*. She has a PhD in Environmental History and Aesthetics. Jessica is the founding editor of *The Willowherb Review* and teaches creative writing at the University of King's College. She lives in Berlin.

Olivia Laing

is a widely acclaimed writer and critic. They're the author of several books, including *The Lonely City*, *Everybody* and *Funny Weather*. Their first novel, *Crudo*, was a *Sunday Times* top-ten bestseller and won the 2019 James Tait Memorial Prize. Their work has been translated into twenty-one languages and in 2018 they were awarded a Windham Campbell Prize for non-fiction.

Joycelyn Longdon

is an award-winning environmental-justice technologist, communicator and PhD Candidate at the University of Cambridge. Her research centres on the design of justice-led conservation technologies for monitoring biodiversity with local forest communities in Ghana, specialising in biodiversity-monitoring through bioacoustics. Joycelyn was 2022's winner of the Emerging Designer London Design Medal, and was featured in British *Vogue*'s December 2023 'Forces for Change' issue. Her debut book, *Natural Connection: What Indigenous Wisdom and Marginalised People Teach us about Environmental Action*, was published in 2025.

Raqs Media Collective

are a group of a media practitioners based in Delhi. Founded in 1992 by Jeebesh Bagchi, Monica Narula and Shuddhabrata Sengupta, they enjoy playing a plurality of roles, usually appearing as artists, sometimes as curators and at other times as philosophical agents provocateurs. Their work has been exhibited at Documenta, the Venice Biennale, Manifesta and the Istanbul, Shanghai, Sydney and Taipei Biennials. Raqs have had solo exhibitions at Tate Exchange, London; Fundacíon Proa, Buenos Aires; Laumeier Sculpture Park, St Louis; Mathaf Museum, Doha; K21, Düsseldorf and the Whitworth Art Gallery, Manchester. Raqs is a word in Persian, Arabic and Urdu for the state that 'whirling dervishes' enter into when they whirl – it embodies the practice of a kinetic contemplation of the world.

Georges Perec (1936–82)

was a French novelist, filmmaker, documentalist and essayist. He won the Prix Renaudot in 1965 for his first novel, *Things: A Story of the Sixties*, and went on to exercise his unrivalled mastery of language in almost every imaginable kind of writing.

Yasmine Hafez

is a doctoral researcher at SOAS, University of London, a Research Fellow at CEDEJ Egypt, and a Research Fellow at the Water Diplomacy Center at Jordan University of Science and Technology. Her thesis, 'Lakeview: An Alternative History of Nile Basin Water Politics', is the result of extensive fieldwork conducted in various fishing hubs around Lake Victoria in Uganda and Kenya and the transboundary wetland in between, as well as in Egypt around the Delta Lakes, namely Lake Burullus, Lake Mariout and Lake Edko. Through her research, she delves into the colonial histories of these lakes, the neoliberal interventions that have shaped these water bodies, and the fishermen's everyday experiences amid the growing threats of climate change. She is also a co-founder of Water Justice for Gaza.

Anthony Acciavatti

works at the intersection of landscape and the history of science and technology. He is the author of the award-winning *Ganges Water Machine: Designing New India's Ancient River*, which is based on walking and boating across the river basin for nearly a decade. In 2023, the Victoria and Albert Museum acquired the instruments and drawings he made of the basin for the permanent collection. His next two books, *Building A Republic of Villages* and *The Values of Imprecision*, will be released in 2026. Acciavatti is the inaugural Diana Balmori Assistant Professor at Yale University and leads Ganges Lab at Collaborative Earth.

Karan Shrestha

is an artist based in Kathmandu and Delhi. His artistic practice includes drawings, sculpture, photographs, films and texts that speak to the complex, entangled relations of Nepal's recent history. Shrestha's works are an archive of the terrain, political histories and transient memories, and a speculative world that suspends reality, to question ideas of progress while drawing connections to the ecological, cultural and socio-economic dimensions of Nepali life. Shrestha has presented work in numerous regional and international institutions.

Gaylene Gould

is a multidisciplinary artist who creates works that unearth buried stories in places, people and cultures, especially those that exist on the margins. Her works are research-led and often participatory, making room for multiple voices alongside her own.

Calthorpe Community Garden

is a community development space which was wrested by the local community from commercial developers in 1984. Calthorpe is led by women 'well-keepers' including Mila Campoy, Annika Miller Jones, Nicole Colombo, Theresa Dauncey and many other community guides and is home to the Black Mary Project, a public-art initiative led by Gaylene Gould, producer Zaynab Bunsie and other Black women artists. The project includes a new permanent memorial by Marcia Bennett-Male, supported by the Mayor of London, memory-sharing workshops, a festival, performances, an installation and a new textile artwork all in honour of Mary Woolaston, the seventeenth-century well-keeper.

Lucille Clifton (1936–2010)

was one of the most distinguished, decorated and beloved poets of her time. She won the National Book Award for Poetry for *Blessing the Boats* and was the first African-American female recipient of the Ruth Lilly Poetry Prize for lifetime achievement from the Poetry Foundation. In 1987 she became the first author to have two books of poetry – *Good Woman* and *Next* – chosen as finalists for the Pulitzer Prize in the same year. She was also the author of eighteen children's books, and in 1984 received the Coretta Scott King Award from the American Library Association for her book *Everett Anderson's Goodbye*.

About the Authors

Lucy Jones

is a writer and journalist based in Hampshire, England. She previously worked at *NME* and the *Daily Telegraph*, and her writing on culture, science and nature has been published in *GQ*, *BBC Wildlife*, the *Sunday Times*, the *Guardian* and the *New Statesman*. She is the author of *Foxes Unearthed*, which won the Society of Authors' Roger Deakin Award 2015; *Losing Eden*, which was longlisted for the Wainwright Prize and named a *Times* and *Telegraph* Book of the Year; and *Matrescence*, 'a thrilling examination of what it means to be a mother' (*Observer*), which has been longlisted for the inaugural Women's Prize for Non-Fiction. She is also the co-author of *The Nature Seed*.

Jalaluddin Rumi (1207–73)

was a poet, Islamic scholar and Sufi mystic originally from the Greater Khorasan in Greater Iran.

Acknowledgements

The extract from 'lake-loop' by Natalie Diaz (copyright © 2020) was co-commissioned by the Academy of American Poets and the New York Philharmonic as part of the *Project 19* initiative and published in Poem-a-Day on 28 March 2020 by the Academy of American Poets.

'The world is blue at its edges and in its depths' is an extract from *A Field Guide to Getting Lost* by Rebecca Solnit, reprinted with permission of Canongate Books through PLSclear.

'Water Justice' is a revised essay from Minority Rights Group's report 'Minority and Indigenous Trends 2023: Focus on Water'.

The extract from Ocean Vuong's poem 'Immigrant Haibun' from *Night Sky with Exit Wounds* was first published by Penguin Random House.

'Ancient Green' is an adapted essay by Robin Wall Kimmerer, first published in *Emergence Magazine*.

A river passing through a landscape' is an extract from *To The River* by Olivia Laing, reprinted with permission of Canongate Books through PLSclear.

'The Parable of the Step Well' by Raqs Media Collective was first published by the World Weather Network for the 'Latitude 28 Degrees North' at Khoj, Delhi.

The extract from *Species of Spaces* by Georges Perec was first published by Penguin Classics.

Lucille Clifton, 'water sign woman' from *How to Carry Water: Selected Poems of Lucille Clifton*. Copyright © 1991 by Lucille Clifton. Reprinted with the permission of the Permissions Company, LLC on behalf of BOA Editions Ltd, boaeditions.org.

Thanks to Dr Alireza Nurbakhsh for his advice on translating Rumi's *Masnavi*.

While every effort has been made to contact copyright-holders of copyright material, the authors and publishers would be grateful for information where they have been unable to trace them and would be glad to make amendments in further editions.

About
the Exhibition

Thirst: In Search of Freshwater is a major exhibition at Wellcome Collection (26 June 2025–1 February 2026), curated by Janice Li.

Thirst is a universal human experience, shared with most other living beings. With only 3 per cent of the Earth's water being fresh, our land thirsts, too. The exhibition explores humanity's vital connection with freshwater, an essential source of life and a pillar of good health for life forms and land masses.

Spanning ancient Mesopotamia and Iran to Victorian London, and extending to modern-day Nepal, Lebanon and Singapore, it combines art, science, history, technology and Indigenous knowledge to deepen the understanding of our relationships with freshwater.

Featuring 125 objects – including thought-provoking artworks, historical artefacts, rich material culture and the latest discovery research – the exhibition immerses visitors in five water conditions: Aridity, Rain, Glaciers, Surface Water and Groundwater. It examines the impact of freshwater access on health and ecosystems, highlights the consequences of mismanagement, such as disease and climate disasters, and presents community-driven, regenerative solutions to the global water crisis.

wellcome collection

Wellcome Collection is a free museum and library. We believe everyone's experience of health matters. Through our collections, exhibitions and events, in books and online, we explore the past, present and future of health. We're part of Wellcome, a charitable foundation supporting science to help build a healthier future for everyone.

wellcomecollection.org